Environmental Justice

What is environmental justice? Why is it important to social work? Social work has a long history with the term "environment" as part of our guiding ecological framework. Yet, frequently the assessment of the environment is left out or seen as a difficult domain to understand as part of the complexities of the human experience, particularly for those who are vulnerable, oppressed and poor. This book consists of ten chapters that showcase environmental justice work in social work practice and in educational settings from around the world. The profession now seems ready to ensure that the "environment" is once again given the consideration it deserves when we are assessing problems in living. These chapters were selected for their helpfulness in moving the profession forward conceptually while offering practical tools for the classroom, for institutions and for further scholarship. This book was originally published as a special issue of *Social Work Education: The International Journal.*

Sondra J. Fogel is an Associate Professor in the School of Social Work and the Director of Special Programs in the Honors College at the University of South Florida, Tampa, USA. She is the current editor-in-chief of *Families in Society: The Journal of Contemporary Social Services.* Her published research areas include homelessness, capital punishment and macro social work practice.

Carenlee Barkdull is an Associate Professor and the Chair of the Department of Social Work at the University of North Dakota, Grand Forks, USA. She has numerous years of experience in the public and nonprofit sectors encompassing policy advocacy, administration and community practice. Several years of research in North Dakota's Bakken oil fields have led to a keen interest in the effects of extractive industries on vulnerable populations and the human service sector.

Bret A. Weber is an Associate Professor in the Department of Social Work at the University of North Dakota, Grand Forks, USA. As a member of the Grand Forks City Council and as co-director of the North Dakota Man Camp Project (including service on the boards of the local housing authority and community land trust), his work focuses on social justice issues related to housing and the social, physical and economic environment.

Environmental Justice

An issue for social work education and practice

Edited by

Sondra J. Fogel, Carenlee Barkdull and Bret A. Weber

Routledge
Taylor & Francis Group
LONDON AND NEW YORK

First published 2016 by Routledge

2 Park Square, Milton Park, Abingdon, Oxfordshire OX14 4RN
711 Third Avenue, New York, NY 10017

Routledge is an imprint of the Taylor & Francis Group, an informa business

First issued in paperback 2018

British Library Cataloguing in Publication Data
A catalogue record for this book is available from the British Library

ISBN 13: 978-1-138-66976-5 (hbk)
ISBN 13: 978-1-138-39209-0 (pbk)

Typeset in Minion
by diacriTech, Chennai

Publisher's Note
The publisher accepts responsibility for any inconsistencies that may have arisen during the conversion of this book from journal articles to book chapters, namely the possible inclusion of journal terminology.

Disclaimer
Every effort has been made to contact copyright holders for their permission to reprint material in this book. The publishers would be grateful to hear from any copyright holder who is not here acknowledged and will undertake to rectify any errors or omissions in future editions of this book.

Contents

Citation Information

The following chapters were originally published in *Social Work Education*, volume 34, issue 5 (August 2015). When citing this material, please use the original page numbering for each article, as follows:

Chapter 1
Rethinking Social Work's Interpretation of 'Environmental Justice': From Local to Global
Dawn Philip and Michael Reisch
Social Work Education, volume 34, issue 5 (August 2015) pp. 471–483

Chapter 2
Environmental Justice and Social Work Education: Social Workers' Professional Perspectives
Ande Nesmith and Nora Smyth
Social Work Education, volume 34, issue 5 (August 2015) pp. 484–501

Chapter 3
Changing Gears: Shifting to an Environmental Perspective in Social Work Education
Mel Gray and John Coates
Social Work Education, volume 34, issue 5 (August 2015) pp. 502–512

Chapter 4
Sustainable Social Work: An Environmental Justice Framework for Social Work Education
Samantha Teixeira and Amy Krings
Social Work Education, volume 34, issue 5 (August 2015) pp. 513–527

Chapter 5
Social Work and Sustainable Social Development: Teaching and Learning Strategies for 'Green Social Work' Curriculum
Julie Drolet, Haorui Wu, Matthew Taylor and Allyson Dennehy
Social Work Education, volume 34, issue 5 (August 2015) pp. 528–543

Chapter 6
Cultivators of Change: Food Justice in Social Work Education
Michelle Lee Kaiser, Sarah Himmelheber, Shari Miller and R. Anna Hayward
Social Work Education, volume 34, issue 5 (August 2015) pp. 544–557

Chapter 7
Exploring the Intersection of Environmental Degradation and Poverty: Environmental Injustice in Nairobi, Kenya
Jennifer L. Willett
Social Work Education, volume 34, issue 5 (August 2015) pp. 558–572

Chapter 8
Transforming Social Work Curricula: Institutional Supports for Promoting Sustainability
Kelly Melekis and Valerie Woodhouse
Social Work Education, volume 34, issue 5 (August 2015) pp. 573–585

Chapter 9
Environmental Sustainability and Social Work: A Rural Australian Evaluation of Incorporating Eco-Social Work in Field Education
Fran Crawford, Savana Sabine Agustine, Leah Earle, Ahmed Bawa Kuyini-Abubakar, Yoni Luxford and Hurriyet Babacan
Social Work Education, volume 34, issue 5 (August 2015) pp. 586–599

The following chapter was originally published in *Social Work Education*, volume 35, issue 1 (February 2016). When citing this material, please use the original page numbering, as follows:

Chapter 10
Reflexive Development: A Model for Helping Social Workers Contribute to a Sustainable Global Future
Eric A. Des Marais, Sarah M. Bexell and Subhasis Bhadra
Social Work Education, volume 35, issue 1 (February 2016) pp. 1–14

For any permission-related enquiries please visit:
http://www.tandfonline.com/page/help/permissions

Notes on Contributors

Savana Sabine Agustine is a Lecturer in Social Work and Human Services at Charles Sturt University, Wagga Wagga, Australia. Her research interests include palliative care, grief and loss, death and dying, and ecological social work. Her work has been published in *Australian Social Work* and *Health and Interprofessional Practice.*

Hurriyet Babacan is the Pro-Vice Chancellor of Torrens University, Australia. Her research interests include gender, cultural diversity, international and community development, policy and governance, social justice, and human rights and rural and regional development. She is currently working on an edited collection entitled *Gender and Policy in the Asia Pacific.*

Carenlee Barkdull is an Associate Professor and the Chair of the Department of Social Work at the University of North Dakota, Grand Forks, USA. She has numerous years of experience in the public and nonprofit sectors encompassing policy advocacy, administration and community practice. Several years of research in North Dakota's Bakken oil fields have led to a keen interest in the effects of extractive industries on vulnerable populations and the human service sector.

Sarah M. Bexell is an Adjunct Professor in the Graduate School of Social Work at the University of Denver, CO, USA. She has been engaged in wildlife conservation, conservation education and humane education for over 20 years. She is the author of *Giant Pandas: Born Survivors* (2012), with Zhang Zhihe.

Subhasis Bhadra is an Assistant Professor in the Department of Social Work at Gautam Buddha University, New Delhi, India.

John Coates is a Professor in the Department of Social Work at St. Thomas University, Fredericton, Canada. He is concerned with the discussion of environmental change within mainstream social work, and the impact that changes in our physical environment will have on our health, lifestyles and relationships. He is the editor of *Spirituality and Social Work* (2007), with John R. Graham and Barbara Swartzentruber.

Fran Crawford is a Professor in Social Work in the School of Health at the University of New England, Armidale, Australia. She draws on extensive "hands-on" knowledge of community development and family interventions in rural and remote locations to develop curricula for social work and other students. Her research and publications have clustered around rural social work, working with aboriginal people, community development, and concepts of evidence-based social work practice and international social work.

Allyson Dennehy is a Research Assistant in the Faculty of Social Work at the University of Calgary, Canada. She is involved in a study that aims to build capacity in the social work profession by integrating questionable development and social development principles.

Eric A. Des Marais is currently completing a Ph.D. in the School of Social Work at the University of Denver, CO, USA. His research focuses on resettlement and natural disasters. He is also a professional in mental health services and has extensive experience in providing counselling and psychotherapy to children, adolescents and adults. He will complete his dissertation in 2016. He is currently living in Japan where he is working as a principal at a private Japanese high school.

Julie Drolet is an Associate Professor in the Faculty of Social Work at the University of Calgary, Canada. She has published extensively in international social work and social development, with a particular focus on gender and development, climate change and disasters, international social protection initiatives, international and Canadian field education, qualitative research, and international migration and Canadian immigration.

Leah Earle is based in the School of Health at the University of New England, Armidale, Australia.

Sondra J. Fogel is an Associate Professor in the School of Social Work and the Director of Special Programs in the Honors College at the University of South Florida, Tampa, USA. She is the current editor-in-chief of *Families in Society: The Journal of Contemporary Social Services*. Her published research areas include homelessness, capital punishment and macrosocial work practice.

Mel Gray is a Professor of Social Work at the University of Newcastle, Australia. Her most recent books include *Decolonizing Social Work* (2013), *Social Work Theories and Methods* (2013), *The New Politics of Social Work* (2013) and *Environmental Social Work* (2013). Her work has contributed to key debates in social work on inter alia social work ethics, evidence-based practice, and international and indigenous social work.

R. Anna Hayward is an Assistant Professor in the School of Social Welfare at Stony Brook University, NY, USA. She is interested in fathers' involvement in children's and family services, juvenile justice and evidence-based interventions. Her work has appeared in the *Journal of Evidence Based Social Work, Social Work Education, Journal of Public Child Welfare* and *Research on Social Work Practice.*

Sarah Himmelheber is an Assistant Professor of Social Work at the University of Wisconsin-Green Bay, USA. She has worked predominantly in mental health and homeless services in community-based settings. Her research agenda includes an ongoing focus on food recovery programming as well as other community-level food security interventions.

Michelle Lee Kaiser is an Assistant Professor in the College of Social Work at Ohio State University, Columbus, USA. Her research focuses on how communities can ensure safe, affordable, healthy food for all residents, and her work related to food access, food mapping, chronic diseases in food pantry clients, community food assessments and food-coping strategies uniquely combines social work, health, sustainable development and geography perspectives.

Amy Krings is an Assistant Professor in the School of Social Work at Loyola University, Chicago, IL, USA. Her research interests include community organization, social movements, urban politics and social justice education. Her work has appeared in journals including *Equity & Excellence in Education* and *Social Work Education.* She previously worked as a community organizer, grant-writer and executive director at a nonprofit organization in Cincinnati, OH, USA, where her work included the management of a community-based campaign to reduce street-level gun violence.

Ahmed Bawa Kuyini-Abubakar is the Course Coordinator for the Social Work program in the School of Health at the University of New England, Armidale, Australia. His research has a broad focus on educational and social inclusion in developing countries, child rights and welfare, refugees and migration, and disability and rehabilitation. He is the author of *Re-Thinking Values in Africa: For Collective Well-Being* (2013) and *Working with Minority Students and Youth in Norway* (2013).

Yoni Luxford is a Senior Lecturer in Nursing in the School of Health at the University of New England, Armidale, Australia. She is registered as a nurse and a midwife, and has qualifications in social work, primary health care and education. She has practical experience in managing projects and developing partnerships between health and education sectors to meet the changing needs of health service provision in Australia.

Kelly Melekis is an Assistant Professor in the Department of Social Work at the University of Vermont, Burlington, USA. Her areas of expertise include gerontology, collaboration, mental health, social justice and social welfare. Her work has appeared in journals, including *Health & Social Work*, the *Journal of Public Health Management & Practice* and the *Journal of Gerontological Social Work.*

Shari Miller is an Associate Professor in the School of Social Work at the University of Georgia, Athens, USA. Her research interests include the professional socialization of social workers, social work education, field education and supervisor training, the culture of the profession, ethical decision-making, critical thinking and ecological issues. Her work has appeared in *Social Work Education* and the *Journal of Social Work Education.*

Ande Nesmith is an Assistant Professor in the School of Social Work at the University of St. Thomas, Saint Paul, MN, USA. His work has been published in *Child & Family Social Work, Child and Adolescent Social Work Journal* and *Children and Youth Services Review.* His research interests include outcomes for adolescents in foster care, environmental justice and the use of technology to support vulnerable youth.

Dawn Philip is a 2015 graduate of the Masters in Social Work programme at the University of Maryland, College Park, USA.

Michael Reisch is a Professor of Social Justice at the University of Maryland, College Park, USA. His most recent books include *Social Policy and Social Justice, The Routledge International Handbook of Social Justice* and *The Handbook of Community Practice.* He has played a leadership role in advocacy, professional and social change organizations, focusing on poverty and inequality, particularly as they affect low-income children and families, welfare recipients, immigrants and refugees, and unemployed or homeless people.

Nora Smyth is currently studying for a Bachelor of Social Work degree at the University of St. Thomas, Saint Paul, MN, USA.

Matthew Taylor was a student research assistant in the Faculty of Social Work at the University of Calgary, Canada and is now working in Edmonton.

Samantha Teixeira is an Assistant Professor of Health and Mental Health in the School of Social Work at Boston College, Chestnut Hill, MA, USA. Her research focuses on how neighbourhood environmental conditions affect youth and how youth can be engaged in creating solutions to environmental problems in their communities. She has published on the topics of place-based, comprehensive community interventions that address neighbourhood environmental disparities, youth-led participatory research and environmental justice interventions and education.

Bret A. Weber is an Associate Professor in the Department of Social Work at the University of North Dakota, Grand Forks, USA. As a member of the Grand Forks City Council and as co-director of the North Dakota Man Camp Project (including service on the boards of the local housing authority and community land trust), his work focuses on social justice issues related to housing and the social, physical and economic environment.

Jennifer L. Willett is Evaluation Manager for the Connecticut Department of Mental Health and Addiction Services, where she has worked on criminal justice and drug abuse, crisis intervention for young adults, and improving life chances for youth and young adults with serious mental conditions. Her work has been published in *Social Work Education, Contemporary Rural Social Work* and the *International Journal of Environmental Sustainability*.

Valerie Woodhouse is studying for a Masters of Social Work at the University of Vermont, Burlington, USA. She is passionate about social justice and has worked as a community organizer for several groups and movements, including eco-justice, affordable housing, LGBTQA+ rights and migrant justice.

Haorui Wu is a postdoctoral scholar in the Faculty of Social Work at the University of Calgary, Canada. His research focuses on post-disaster reconstruction from the perspective of social issues, and looks at community planning practices for sustainable development and community resilience post-disaster amongst children and youth.

Introduction: Environmental Justice and Social Work

Sondra J. Fogel, Carenlee Barkdull & Bret A. Weber

Social workers know that individuals, communities and the larger society must be understood within a broad environmental context. This perspective encompasses all the diverse aspects of social work practice and distinguishes it from other helping professions. This is particularly significant in the face of ongoing environmental and climatic shifts that disproportionately threaten vulnerable populations around the world—a matter of global concern requiring interventions at all societal levels. With specific training, ethical codes and unique perspectives directly aimed at enhancing individual and societal well-being, more than any other profession, social work is uniquely positioned to address this critically important set of related issues. And yet, as noted by Zapf (2009), social work practice has "been strangely silent" about the natural, physical environment and the related "threats to human well-being and continued existence" (p. 18). This book helps to fill the current need for additional content on environmental justice issues specifically for social work in order to help faculty and instructors prepare students and future leaders in the field to confront the increasing environmental threats experienced by individuals, other living species and communities worldwide.

Tackling environmental injustices will involve confronting issues of power and privilege. However, this is a challenge that we must address. The wealthiest populations are the greatest consumers of the products and activities damaging to the global environment, yet the impoverished are most likely to suffer the consequences. For instance, the United States is the world's top emitter of carbon dioxide, but the cost of climate change is disproportionately borne by those least able to either afford the costs or recover from the effects. As noted in *Global Climate Risk Index 2011* in relation to recent extreme weather events, "all of the ten most affected countries (1990–2009) were developing countries in the low-income or lower-middle income country group" (Harmeling, 2010, p. 5). For example, one related consequence is that climate change and rising sea levels will displace tens of millions of Bangladeshis over the coming decades, and their agriculture is already suffering from the salinization contaminating delta regions (Chopra, 2009). Throughout the developing world, social workers tend to focus on social development and issues of sustainability. Increasingly, those efforts are directly related to environmental issues including population health risks, decreased food supplies and disasters connected to natural devastation.

Meanwhile, social work in the United States (with some notable exceptions) has largely been silent on these issues. That disregard has been noted along with the beginning of a groundswell of understanding of environmental injustices and a willingness to take action. Of particular note, in the United States, the Council on Social Work Education in its 2015 revised EPAS (Educational Policy and Accreditation Standards) explicitly used the term "Environmental Justice" in the title of its third competency. The change in US accreditation standards should not be a surprise, as there is a broad understanding across all forms and levels of practice that systems operate in complex, interdependent ways. And yet, there is a persistent tendency—particularly in the United States—to provide a nod to the physical environment, and to overemphasize the internal environment of the mind, underscoring the conceptual and practical challenges to address environmental justice in social work practice and education. To fulfill its mission, social work must engage with the rapidly changing physical environment, and to do so it requires transformation in educational content.

Rather than as an emergent issue in social work, it is perhaps better to consider the broader "environment" in terms of a long-overdue reconnection. As a central pioneer in the emergence of professional social work, Jane Addams' role as one of the most important, modern urban planners is insufficiently recognized by many of today's social workers. Throughout her career, she emphasized the connection between individual well-being and physical conditions, consistently returning to matters of public health, housing, the cityscape and the propinquity of all these matters with national and international policies. It is not difficult to imagine that if Addams were practicing today, environmental issues would be central to her agenda.

However, it is not as though nothing has happened between the Progressive Era in which Addams practiced and the present. An active intersection with policy initiatives began gaining momentum by the 1970s, followed soon after with the 1980s' emergence of an environmental justice movement. More directly specific to social work, Marie D. Hoff and John G. McNutt's (1994) book sought to capture the contributions, already evident by that time, of social workers addressing environmental issues. More recently, Michael Zapf's *Social Work and the Environment* (2009) proposed a theoretical shift to "people and place" as a way to address the profession's neglect of the physical environment, and Lena Dominelli's *Green Social Work* (2012) articulates and advances the possibility of a new practice paradigm to affirm human rights and address structural inequalities amidst environmental degradation caused by neoliberal policies. Of course, these few references represent only a sample of the emergent recognition about environmental injustice issues around the globe and the related ethical obligations for social work. This book seeks to build on those contributions as another key step in the process to harness the energy of the profession to assess and address environmental threats in all practice and policy interventions.

The included chapters describe the concentrated efforts by individual faculty members and advanced students who are striving to infuse ecological content into practice courses, new concentrations and university programs. Social workers everywhere should be encouraged by the outstanding work occurring in the classroom and field. These selected works were chosen for their helpfulness in moving the field forward conceptually while simultaneously offering practical tools for the classroom, institutions and scholarship.

The first chapter exposes some of the deficiencies in current social work education in the United States. Chapter 1 provides the historical grounding, particularly in the United States, where the profession—as has been noted—lags behind its global counterparts. The differences between environmentalism, environmental justice and environmental racism are illustrated in historical context. This context is further illuminated by a case study of a social work student in an urban setting, with lessons drawing from the use of transformational learning theory and green social work in the classroom.

The emergence of research connecting the social work profession to environmental issues signals both the growing awareness and the potential for disciplinary transformation. Chapter 2 shares the results of a study of 373 US-based social workers who acknowledge deficiencies in their awareness of these issues and the related impacts on client populations. Most significantly for future trends, respondents express their dissatisfaction with their educational preparation to address these issues.

The remaining chapters provide practical ideas and tools for transforming social work education to be more actively connected to ecological concerns. Both Chapters 3 and 4 delve deeply into social work's theoretical frameworks and perspectives to help educators grapple with the underlying meanings and assumptions inherent in the manner in which these theories are currently taught. Both chapters argue that social work's mission, values and practice with society's marginalized, oppressed and disadvantaged populations make engagement with environmental issues both necessary and morally inevitable. These works demonstrate how key concepts can be utilized in the classroom to help students make theory-to-practice connections in ways that promise deeper integration of environmental concerns into emergent professional identities.

Chapter 5 provides a focus solely on the classroom, including practical tools and examples from an elective course on social work and sustainability. Focusing on undergraduate learners, it provides a potential resource for social work educators seeking to equip budding social workers with the concepts and methods to engage environmental issues at all levels of practice.

Chapter 6 also focuses on the student learner, broadening the view from one course to share examples of assignments, units and courses from other disciplines to help draw multidisciplinary connections to the issues of food security, food justice and environmental justice in social work. The case example presented by Jam City in Nairobi, Kenya, in Chapter 7 helps students understand the value of structural analysis for unveiling and illuminating the causes and consequences of environmental injustice in a powerful set of examples further demonstrating the impacts of neoliberal policies from the global to the local. While this case study is set in a developing country, implications for social work education in developed countries are also explored.

Finally, similarly pragmatic tools for social work education administrators and faculty can be found in Chapters 8 to 10. The first chapter mentioned seeks to deepen the engagement of faculty leaders within organizational units in which social work is housed, and provides tools for building cross-disciplinary supports that deepen broader institutional commitment. The next chapter shares a survey conducted with field faculty and field supervisors, which illuminates possibilities and challenges of translating environmental competencies from classroom to practice. The final chapter provides helpful classroom activities based on a reflexive development model to help social workers understand global social ecological systems and act in accordance to environmental justice and sustainable practice principals.

This book arrives at a fairly nascent period of global social work discussion on this topic, and there remains much to be done to embrace this concept within social work education, as Jeffery (2014) points out. This book is meant to serve practitioners in the field, faculty and instructors looking for ways to incorporate more environmental content into their courses, and students wanting to gain a broader understanding of this issue. Most importantly, it is our hope that this book will contribute to a growing movement in our profession to embrace environmental justice, not as a broad, nebulous concept, but as a meaningful manifesto that inspires a growing number of social workers to actively embrace their ethical obligations as change agents who address the causes and consequences of environmental degradation on our planet.

References

Chopra, A. (2009, March 26). How global warming threatens millions in Bangladesh. *U.S. News*. Retrieved on February 24, 2016, from www.usnews.com/news/energy/articles/2009/03/26/how-global-warming-threatens-millions-in-bangladesh.

Dominelli, L. (2012). *Green Social Work: From Environmental Crises to Environmental Justice*. Cambridge, UK: Polity.

Harmeling, S. (2010, December). *Global Climate Risk Index 2011: Who suffers most from extreme weather events? Weather-related loss events in 2009 and 1990 to 2009*. Retrieved on August 25, 2011, from www.germanwatch.org/klima/cri2011.pdf.

Hoff, M. D., & McNutt, J. G. (Eds.) (1994). *The Global Environmental Crisis: Implications for Social Welfare and Social Work*. Brookfield, VT: Avebury.

Jeffery, D. (2014). Environmentalism in social work: What shall we teach? *Affilia: Journal of Women and Social Work, 29*(4), 492–498.

Zapf, M. K. (2009). *Social Work and the Environment: Understanding People and Place*. Toronto, Ontario: Canadian Scholar's Press, Inc.

Rethinking Social Work's Interpretation of 'Environmental Justice': From Local to Global

Dawn Philip & Michael Reisch

This article challenges social workers to expand their understanding of the 'person-in-environment' perspective and become more active in addressing current environmental crises. Although social work scholars have begun to explore the relationship between social work and the natural and built environment and professional organizations mandate the integration of this content into practice and education, these goals remain unrealized, particularly in the USA. To address these issues more effectively, social work educators will need to distinguish between understanding persons in their environment and environmentalism, and between environmentalism and environmental justice. This article analyzes the emergence of the environmental justice movement in the USA and other nations and its relationship to environmental racism. It presents a case study of a local environmental justice effort to demonstrate how social workers can use their knowledge and skills to make important contributions to environmental justice and sustainability. It also discusses the potential of 'green social work' and transformative learning theory as tools to help social work educators better equip students to make strategic alliances across professions, disciplines, and systems to address contemporary environmental crises.

Introduction

Scholars have recently begun to explore the relationship between social work and the natural and built environment (Gray & Coates, 2013; Pulla, 2013); and professional organizations (National Association of Social Workers, 2009; International Federation of Social Workers, 2013) mandate the integration of this content into practice and education. Yet, to a considerable extent, these goals remain unrealized, particularly in the USA, although social justice principles and the person-in-environment perspective have formed the foundation of the profession for over a century. Because of the primacy of selfhood over collective rights, contemporary social work practice in the USA focuses more on adaptation than on challenging the structures underlying environmental inequities (Reisch & Jani, 2012). Consequently, the relationship between persons and the environment is viewed in a static, individually oriented manner that fails to recognize the differences that exist, among and within nations, based on cultural norms and political-economic realities (Jani & Reisch, 2011).

These differences appear in accounts of creation itself. Many indigenous narratives emphasize harmony between humans and the Earth. By contrast, Western religious and secular ideas about this relationship reflect an anthropocentric view of the universe that rationalizes environmental exploitation and destruction in the name of progress.[1]

The earliest Western concerns about the environment emerged from both idealistic and scientific perspectives. In response to the social impact of the Industrial Revolution, the nineteenth-century Romantic Movement celebrated a pre-industrial view of nature, while the public health movement responded to increased evidence that environmental degradation had deleterious health and social effects, particularly on the urban working class.[2] Yet, the US government did not overcome corporate opposition and develop major environmental initiatives until the 1960s and 1970s, inspired in part by Rachel Carson's book, *Silent Spring* (1962). It soon became clear, however, that these reforms did not address the environmental injustices that disproportionately affected racial and ethnic communities, indigenous peoples, and low-income persons all over the world (Bullard, 1994).

This article challenges social workers to expand their understanding of the 'person-in-environment' perspective and become more active in addressing current environmental crises. To address these issues effectively, social work educators will need to articulate the links between the 'person-in-environment' perspective and environmentalism and distinguish between environmentalism and environmental justice. We assert that social work's 'ambivalent understanding of its relationship to the natural world' is not only unacceptable, but arguably unethical (Besthorn & Saleebey, 2003) and suggest how educators can better equip students to make strategic alliances across professions, disciplines, and systems to address these crises. After summarizing the evolution of social work perspectives on the environment in the USA, we discuss the emergence of the environmental justice movement and its relationship to environmental racism, describe how globalization influences international perspectives on the environment, and, through a case study, outline how social workers can use their knowledge and skills to make important contributions to environmental justice and sustainability.

Social Work and the Environment

The responses of the social work profession to environmental issues paralleled those of US society as a whole. In the late nineteenth and early twentieth centuries, social workers, particularly in the settlement house movement, were at the forefront of the public health response to urban problems created by industrialization. They recognized the relationship between the physical environment and human well-being, that '*where* people live profoundly influences *how* they live, with important implications for ... social justice' (Kemp, 2011, p. 1200). By addressing these public health and psycho-social effects, radical social workers articulated an early version of environmental justice that resonates to some extent today in the work of the Social Work Action Network (SWAN) in the UK and the Social Welfare Action Alliance (SWAA) in the USA (Reisch & Andrews, 2002).

During the twentieth century, this emphasis declined in US social work for several reasons. In pursuit of professional status, social workers increasingly focused on individual or family problems rather than structural issues (Specht & Courtenay, 1994). Many social workers embraced the popular view that economic growth, rather than redistribution, would ultimately reduce the environmental injustices poverty created. They viewed the solution to environmental injustice in terms of adaptation to environmental conditions or amelioration of the problems these injustices produced, rather than eradicating their underlying causes (Reisch & Andrews, 2002).

Although Richmond (1917) exhorted social workers to 'treat individuals by way of their social environment,' and social work theory consistently emphasizes an ecological, person-in-environment approach (Karls & Wandrei, 1994), practitioners tend to regard the environment as the *context* for practice instead of a dynamic component of people's lives (Jani & Reisch, 2011). The recent focus on 'evidence-based practice' has transformed social work research from analysis of structural causes of societal inequities to measurement of the effectiveness of scientifically determined 'interventions,' tacitly accepting the status quo (Reisch & Jani, 2012).

The Environmental Justice Movement and Environmental Racism

The idea that environmental forces determine the quality of human well-being aligns with key principles of the contemporary environmental justice movement, which emerged during the early 1980s, fueled by growing concerns that toxic facilities were disproportionately located in low-income communities of color. Described as a 'marriage of the movement for social justice with environmentalism' (McGurty, 1997, p. 303), the environmental justice movement applied a civil rights approach and inspired a broad range of local struggles to redress existing environmental inequities through targeted action. Unlike frameworks that stress the preservation of ecosystem biodiversity and natural resources, an environmental justice framework centers the experiences of populations who are most affected by environmental degradation and enhances their ability to participate meaningfully in local environmental decision-making (Bullard, 1994). Developing robust public participation mechanisms, for example, allows communities to proactively challenge inequitable environmental decisions made on their behalf. Many barriers to such participation still exist, including limited access to health and environmental data and insufficient funding for technical assistance that would facilitate a community's understanding of the often complex scientific and regulatory policies that drive environmental decision-making. Reducing these barriers is critical if low-income communities are to have a voice in shaping the environmental decisions that directly affect their short- and long-term health and well-being.

A statement by the US Environmental Protection Agency (United States Environmental Protection Agency, 2014) acknowledges this imperative. It defines environmental justice as the fair treatment and meaningful involvement of all people regardless of race, color, national origin, or income with respect to the development, implementation, and enforcement of environmental law, regulations, and policies.

At the local level, environmental justice can be advanced by analyzing the consequences of undesirable land use (Maller, Townsend, Pryor, Brown, & St. Leger, 2006). For example, there has been increased attention recently to the impact of hydraulic 'fracking' by extractive industries on racial and ethnic minority, low-income, and rural communities in the USA and developing nations (Davis, 2012; Kovats et al., 2014).

The environmental justice movement also inspired the term 'environmental racism,' which Bullard (1993–1994) defined as 'any policy, practice or directive that differentially affects or disadvantages (whether intended or unintended) individuals, groups, or communities based on race or color' (p. 1037). This term gained currency after the United Church of Christ Commission for Racial Justice (1987) issued a groundbreaking report demonstrating how race, more than any other factor, predicted the location of toxic waste facilities. There is abundant evidence of this phenomenon. The impact of uranium mining in the North American West on Indigenous people; the frequent exposure of Latino farm workers and Indian cotton farmers to toxic pesticides; and the effects of illegally dumped manufacturing waste on the drinking water of residents along the USA-Mexico border are just a few examples.

Lack of Diversity Within Mainstream Environmental Movements

Environmental racism is also reflected in the composition of the environmental movement. During the first half of the twentieth century, environmentalists focused primarily on wilderness and wildlife protection (Worster, 2008). These goals mirrored the interests of the movement's initial supporters—privileged white populations who wanted to retain their ability to enjoy outdoor recreational opportunities, particularly in the nation's new national parks. Early environmental organizations excluded racial minorities and working-class people and largely ignored the environmental problems of inner cities, because they regarded urban areas as places of physical degradation and moral decay.

Today, many mainstream environmental organizations still reflect these initial characteristics. Consequently, many low-income and minority communities feel alienated by the environmental movement and consider it elitist. They view the reforms it promotes as disproportionately benefiting the movement's supporters while imposing costs on them (Bullard, 2005).

In response to this neglect, the environmental justice movement gained momentum and relevance. In 1991, more than 650 grassroots and national leaders representing over 300 environmental groups attended the First National People of Color

Environmental Leadership Summit. In 1994, President Clinton issued Executive Order 12898, the first major federal initiative requiring all government agencies to collect data on the health and environmental impact of their programs on minority and low-income populations (Executive Order No. 12898, 1994). By the end of Clinton's second term, however, resistance from Congress, industry groups, and even within the EPA, significantly blunted this push for environmental justice. Today, however, environmental advocates are making important progress in preventing undesirable and unaccountable local land use and other environmental injustices in low-income communities through targeted policy reforms, public education campaigns, and sustained community organizing. The case study below illustrates how social workers can be important catalysts in advancing these efforts.

Case Study

In September 2009, Energy Answers International (EAI) applied to construct the largest waste-to-energy (WTE) incinerator in the USA in the Curtis Bay neighborhood of Baltimore. The incinerator would be less than a mile from a high school; over 230 trucks each day would transport 4,000 tons of waste through Curtis Bay and surrounding communities (Smith, 2014). Curtis Bay is already overburdened by toxic facilities, including one of the largest medical waste incinerators in the country, chemical plants, fuel depots, and a 200-acre coal pier (Williams, 2015). It has some of the poorest health statistics in Maryland and the dubious distinction of ranking first in the USA for its concentration and quantity of toxic air pollutants (Massachusetts Institute of Technology Laboratory for Aviation and the Environment, 2013). Given this overall community profile, a cumulative risk assessment would likely demonstrate the compounding effects of additional noxious facilities (Sexton & Linder, 2010). Although the WTE industry insists that incineration is environmentally and economically efficient, environmental advocates and the United Nations have pointed out the numerous adverse effects of incinerators on human health, climate change, and efforts to reduce waste and promote recycling.

Inspired by the incinerator project, a group of local high-school students initiated a campaign to block construction, using creative strategies to highlight community opposition to the project and undermine the incinerator's financing plan. Greg Sawtell, a social worker and organizer with United Workers, a Baltimore-based human rights group that advocates for low-wage workers, has worked with these students for several years on an innovative human rights study project called Free Your Voice (FYV). They built a small, but growing movement to stop construction and raise the environmental literacy, social justice consciousness, and leadership skills of these youth. The students' campaign stresses the connection between human rights and the impact of recent developments on their communities. It addresses such questions as 'What broader structural changes are needed to facilitate our human rights?'

FYV investigated how local and state authorities allowed EAI to construct the incinerator in an environmentally overburdened community. Students learned that only a handful of community members knew about the project, although many residents had serious concerns about the health risks associated with living in such close proximity to a large incinerator. To increase community awareness, Sawtell worked with FYV to attract media attention through such tactics as a march from the school to the proposed incinerator site, a presentation to the Baltimore City School Board using rap performances and poetry reading, and testimony before the state legislature supporting a bill mandating a cumulative health risk assessment when siting environmentally hazardous facilities (Smith, 2014).

Another effective tactic targeted EAI's energy-buying contract with local government entities. The students informed the School Board that it could opt out of its contract to purchase power from EAI if the plant did not deliver electricity within 48 months of the agreement (April 2015). For a number of reasons, including increased local opposition, the project's start date has been delayed until at least 2016. As a follow-up, FYV addressed Baltimore City's Commission on Sustainability. The group is now exploring zero-waste strategies that would create local recycling and composting jobs, strengthen the community's economic base, and obviate the need for an incinerator. Sawtell's success in building community resistance to environmental injustice illustrates how social workers can amplify the voices of marginalized constituencies in creating a more environmentally just world.

Globalization and Environmental Justice

Despite important environmental justice victories highlighted by the case study above, during the past several decades, economic globalization has been a major obstacle to the development of a robust global environmental justice agenda. Hancock (2003) claims that global capitalism subjugates human rights and environmental protection to economic growth and the maintenance of inefficient patterns of production, while undermining the ability of communities to extract environmental concessions from corporations that seek to locate or expand in their area. Communities are thus compelled to sacrifice environmental quality, occupational safety, and political control in the pursuit of illusory economic gains. Current policy-making mechanisms remain locked in anachronistic patterns, while US courts and international institutions, such as the World Trade Organization (WTO), have invalidated environmental, health, and worker safety laws and shielded corporations from any environmental liability. The combination of institutional failures and framing the issue as 'environmental quality vs. growth' has permitted corporations to exploit natural resources in the most profitable and least sustainable ways and transfer social costs onto vulnerable populations in such diverse places as Central Europe, China, Brazil, sub-Saharan Africa, and the USA (Ife, 2008).

In response to these developments, environmental justice activists have emphasized four major themes (Kuehn, 2014): equitable distribution of environmental risks and benefits (Korgen, White, & White, 2011); fair and meaningful participation in environmental decision-making; recognition and appreciation of diverse ways of life, local knowledge, and cultural differences; and the capability of communities and individuals to function and flourish in their societies (Schlosberg, 2007). For example, the environmental justice movement in the UK stresses the relationship between poverty and the environment, and health inequalities and social exclusion. The South African Constitution (Republic of South Africa, 1996) grants citizens the right to an 'environment that is not harmful to their health or well-being' and 'to ... secure ecologically sustainable development and use of natural resources. ' Environmental justice activists in India link its attainment to human rights and access to adequate environmental resources for a healthy life; emphasize its importance for the poorest and least powerful people; regard environmental justice as an intergenerational issue; and recognize that the issue cannot be left to environmental activists alone (DYUTI, 2011).

Despite these statements of principle, the creation of an effective cross-national strategy to counteract globalization's far-reaching negative effects on low-income communities will require the development of sustainable international coalitions among groups that have a long history of antagonistic relationships or that are largely ignorant of each other's existence: international trade unions, environmental groups, civil, women's and human rights organizations, transnational NGOs, academic researchers, and some liberal economists. While many US-based environmental justice networks and coalitions now include international partners, with a few notable exceptions, US social workers have not been active in anti-globalization or environmental justice efforts (Shaw, 2013; Schmitz, Matyók, Sloan, & James, 2012).

Nevertheless, during the past several decades, social workers have shifted from a primarily local orientation towards a more global perspective that highlights the importance of human rights and environmental sustainability (Healy, 2001). Although some scholars and activists have criticized the human rights framework as individually focused and Western-centric (Ife, 2007), it underscores the universal entitlement to the basic environmental necessities of life. Yet, while there is extensive literature examining the connections between human rights and social work (Ife, 2008; Reichert, 2011) and between international social work and environmentalism (Zapf, 2009), there is scant mention of environmental justice within mainstream social work literature (Hawkins, 2010) or, except among radical social workers, of the disproportionate cost that the world's poor pay for ecological damage (Sachs, 2009). Despite their widespread provision of direct services to victims of environmental crises, most social workers are largely absent from the policy-making structures that shape the quality and breadth of these services and help determine how future environmental crises can be prevented (Coates, 2003).

Dominelli's concept of 'green social work' presents a way for social workers to fill this gap. It focuses on how responses to environmental crises must occur not only through service delivery, but also by proactively addressing socio-economic disparities, consumption patterns, limited resources, and the manifest connections between poverty and the absence of environmental rights (Dominelli, 2012). As a conceptual framework, green social work helps clarify the global dimensions of various ecological crises and their impacts on social work's professional ambit. The integration of green social work within anti-oppressive and radical social work traditions can highlight core environmental justice principles, such as dismantling the structural inequalities that expose low-income communities to disproportionate environmental harm. As a practical framework, the concept is useful in enabling social workers to translate their 'on the ground experiences' into concrete efforts that address these structural inequalities.

Dominelli asserts that in order to confront current environmental crises effectively, social workers must apply a sophisticated array of knowledge and skills. As the case study demonstrates, social workers are well-equipped to deploy both clinical and macro-level skills and experience to develop dynamic, culturally relevant responses to these issues. We possess intimate knowledge of the effects of environmental degradation on vulnerable populations that can enhance the representation of these populations in each phase of critical decision-making. We also have an ethical responsibility to use our expertise to go beyond ameliorating the consequences of environmental injustices by preventing further environmental degradation and promoting environmentally sustainable ways of living.

Potential roles include:

- Working with environmentally overburdened communities, urban planners, and public health professionals to create 'bottom-up' ways of eliciting local knowledge through spatial mapping technologies that can be translated to effective interventions (e.g., mapping of food deserts or accessible green space in low-income communities) (Kemp, 2011).
- Using multi-level interventions to articulate essential links between individual well-being, the physical environment, and collective action.
- Facilitating and mediating solution-oriented discussions between stakeholders to develop viable alternatives to environmental projects that prioritize environmental sustainability over profits.
- Developing creative ways to incorporate community-based participatory research in development and planning processes, including environmental impact studies and community health profiles.
- Translating individual and community stories of disproportionate exposure to environmental harms into policies that require cumulative risk assessments and enhanced public participation mechanisms early in the environmental decision-making process.
- Mobilizing communities to engage in collective problem-solving by identifying community assets and power relationships between community members and groups (Dominelli, 2012).

Implications for Social Work Education

These potential practice roles have significant implications for the future of social work education. To better prepare students to address current environmental challenges, social work educators—particularly in the USA—will need to examine long-standing assumptions about practice, such as valuing individual autonomy over community rights and collective responsibility, and expand the meaning of 'person-in-environment' to include the natural and built environment. Recent research indicates that students are receptive to such changes. Although they are introduced to the person-in-environment framework early in their studies, students receive little content on the role of the physical environment and even less on the relationships among human activity, structural inequality, and the natural world. While some educators suggest that additional content on the relationship between social work and environmental degradation may provide a clearer path for students to engage in environmental advocacy (Miller & Hayward, 2014), Jones (2010) asserts that simply adding courses to the curriculum will not produce the fundamental shifts needed to address current ecological crises effectively. He and Mezirow (2003) suggest that transformative learning theory might be an effective way to change how students perceive the world by altering fixed assumptions into ideas that are more flexible and reflective.

At its core, transformative learning theory seeks to shift the ways students see and make meaning of the world by revising problematic frames of reference—habits of mind, fixed assumptions, expectations—to make them more critical, reflective, and ultimately, open to alternative ways of thinking. Encouraging students to reevaluate their relationship to the nonhuman world will first require educators to interrogate and problematize the existing level of ecological alienation prevalent among US social work students. Critical reflection, already a core component of social work education, can be used to challenge students' blind spots in their own knowledge about the nonhuman world, cultural values, and larger economic processes like globalization. Students, for example, could be assigned to write an autobiographical essay discussing how modernity affects their relationship with the natural world or be introduced to a 'disorienting dilemma' that exposes their alienation from the environment. Using class time to visit community gardens, local trash incinerators, or degraded waterways is another instructive way to expose students to the impacts of human behavior on the natural world (Jones, 2010). These educational tools allow social work educators to model eco-oriented social work practice and encourage students to critically examine how larger economic and political forces shape the values and beliefs that contribute to ongoing ecological crises.

In addition, US social work educators could revise how they integrate a global perspective into their teaching (Miller, Hayward, & Shaw, 2012). Although schools have attempted to 'internationalize' their curricula since the 1980s (Link & Ramanathan, 2011), US social workers continue to be relatively ignorant about international human rights norms and their legal enforcement (Ogden, 2008). To be relevant in an increasingly interconnected world, today's students must not only have a more global orientation, they must also recognize the consequences of the Global North's disproportionate consumption and exploitation of natural resources on developing nations (Dominelli, 2012). This includes understanding how untrammeled development and material consumption often contradict the profession's core values (Klein, 2014).

Finally, social work educators could enhance students' understanding of environmental justice by exploring the connections between environmental racism and increasing inequality, environmental degradation, and a general devaluing of the inherent dignity of persons. Expanding the knowledge base of social work practice to include more content on human rights, environmental justice, and sustainable development is a critical first step. It will provide 'social work ... [with] ... the opportunity to be part of the solution rather than an uninvolved bystander to ... emerging environmental predicaments' (McKinnon, 2008, p. 266). A second step would involve discussions of practical interventions, at the individual and community level, to alleviate the consequences of poverty without damaging the environment (Dominelli, 2012).

Conclusion

In order for social workers to integrate environmental concerns into their understanding of social justice, the profession must reconcile the collective responsibility this requires with its longstanding belief in individual self-determination. This is no easy task, either intellectually or practically. Environmental justice is a precondition for all people to exercise agency and improve their individual and collective well-being in a sustainable manner—goals that are certainly compatible with social work's mission and values.

In the future, social work educators and scholars must grapple with such questions as: How can community practitioners enable people to understand the complementary relationship between their individual interests and those of the broader society? How can clinical social workers incorporate environmental concerns into their services while resolving the tension between individual autonomy and collective well-being? A possible solution lies in acknowledging that all people need and want a sense of place and belonging and that these desires are best realized by integrating a respect for the natural environment into every aspect of their lives (Dominelli, 2012).

Disclosure statement

No potential conflict of interest was reported by the authors.

Notes

[1] See, for example, the Greek myth of Prometheus, considered a martyr for giving fire to humankind, and the frequent statements in *Genesis* giving humans dominion over the Earth.

[2] Ironically, Darwin's *Origin of Species* (1859/1956), which placed humans at the top of the evolutionary hierarchy, also introduced the concept of ecology.

References

Besthorn, F. H., & Saleebey, D. (2003). Nature, genetics and the biophilia connection: Exploring linkages with social work values and practice. *Advances in Social Work*, *4*(1), 1–18.

Bullard, R. D. (1993–1994). Environmental racism and invisible communities. *West Virginia Law Review*, *96*, 1037–1050.

Bullard, R. D. (Ed.). (1994). *Unequal protection: Environmental justice and communities of color.* San Francisco, CA: Sierra Club Books.

Bullard, R. D. (Ed.). (2005). *The quest for environmental justice: Human rights and the politics of pollution*. San Francisco, CA: Sierra Club Books.

Carson, R. (1962). *Silent spring*. Boston, MA: Houghlin Mifflin.

Coates, J. (2003). *Ecology and social work: Towards a new paradigm*. Halifax: Fernwood Publications.

Commission for Racial Justice. (1987). *Toxic wastes and race in the United States*. New York, NY: United Church of Christ.

Darwin, C. (1859/1956). *The origin of species*. New York, NY: E.P. Dutton.

Davis, C. (2012). The politics of fracking: Regulating natural gas drilling practices in Colorado and Texas. *Review of Policy Research*, *29*, 177–191. doi:10.1111/j.1541-1338.2011.00547.x

Dominelli, L. (2012). *Green social work: From environmental crises to environmental justice*. Cambridge: Polity Press.

DYUTI. (2011). *Statement at international conference on social work and environmental justice*. Kochi, India: Rajagiri College of Social Sciences.

Executive Order No. 12898. Federal Register 59(32): 7629–33 (1994). *Reprinted in Federal Register*. Washington, DC: U.S. Government Printing Office.

Gray, M., & Coates, J. (2013). Changing values and valuing change: Toward an ecospiritual perspective in social work. *International Social Work*, *56*, 356–368. doi:10.1177/0020872812474009

Hancock, J. (2003). *Environmental human rights: Power, ethics, and law*. London: Ashgate.

Hawkins, C. (2010). Sustainability, human rights, and environmental justice: Critical connections for contemporary social work. *Critical Social Work*, *11*(3): 68–81.

Healy, L. M. (2001). *International social work: Professional action in an interdependent world*. New York, NY: Oxford University Press.

Ife, J. (2007). Cultural relativism and community activism. In E. Reichert (Ed.), *Challenges in human rights: A social work perspective* (pp. 76–96). New York, NY: Columbia University Press.

Ife, J. (2008). *Human rights and social work: Towards rights-based practice* (Revised ed.). New York, NY: Cambridge University Press.

International Federation of Social Workers. (2013). *Integrating human development and environmental sustainability in a new global agenda*. Retrieved from http://ifsw.org

Jani, J. S., & Reisch, M. (2011). Common human needs, uncommon solutions: Applying a critical framework to perspectives on human behavior. *Families in Society: The Journal of Contemporary Social Services*, *92*, 13–20. doi:10.1606/1044-3894.4065

Jones, P. (2010). Responding to the ecological crisis: Transformative pathways for social work education. *Journal of Social Work Education*, *46*, 67–84. doi:10.5175/JSWE.2010.200800073

Karls, J. M., & Wandrei, K. E. (Eds.). (1994). *Person-in-environment: The PIE classification system for social functioning problems*. Washington, DC: NASW Press.

Kemp, S. P. (2011). Recentring environment in social work practice: Necessity, opportunity, challenge. *British Journal of Social Work*, *41*, 1198–1210. doi:10.1093/bjsw/bcr119

Klein, N. (2014). *This changes everything: Capitalism vs. the climate*. New York, NY: Simon & Schuster.

Korgen, K. O., White, J. M., & White, S. K. (Eds.). (2011). *Sociologists in action: Sociology, social change, and social justice*. Thousand Oaks, CA: Pine Forge Press.

Kovats, S., Depledge, M., Haines, A., Fleming, L. E., Wilkinson, P., Shonkoff, S. B., & Scovronick, N. (2014). The health implications of fracking. *The Lancet*, *383*, 757–758. doi:10.1016/S0140-6736(13)62700-2

Kuehn, R. (2014). Environmental justice. In M. Reisch (Ed.), *The Routledge international handbook of social justice* (pp. 319–338). London: Routledge.

Link, R. J., & Ramanathan, C. S. (2011). *Human behavior in a just world: Reaching for common ground*. Lanham, MD: Rowman & Littlefield.

Maller, C., Townsend, M., Pryor, A., Brown, P., & St. Leger, L. (2006). Healthy nature healthy people: 'Contact with nature' as an upstream health promotion intervention for populations. *Health Promotion International, 21*, 45–54. doi:10.1093/heapro/dai032

Massachusetts Institute of Technology Laboratory for Aviation and the Environment. (2013). *Air pollution causes 200,000 early deaths each year in the U.S.* Retrieved from http://lae.mit.edu/air-pollution-causes-200000-early-deaths-each-year-in-the-u-s/

McGurty, E. M. (1997). From NIMBY to civil rights: The origins of the environmental justice movement. *Environmental History, 2*, 301–323. doi:10.2307/3985352

McKinnon, J. (2008). Exploring the nexus between social work and the environment. *Australian Social Work, 61*, 256–268. doi:10.1080/03124070802178275

Mezirow, J. (2003). Transformative learning as discourse. *Journal of Transformative Education, 1*, 58–63. doi:10.1177/1541344603252172

Miller, S. E., & Hayward, R. A. (2014). Social work education's role in addressing people and a planet at risk. *Social Work Education, 33*, 280–295. doi:10.1080/02615479.2013.805192

Miller, S. E., Hayward, R. A., & Shaw, T. V. (2012). Environmental shifts for social work: A principles approach. *International Journal of Social Welfare, 21*, 270–277. doi:10.1111/j.1468-2397.2011.00848.x

National Association of Social Workers. (2009). *Social work speaks: National Association of Social Workers' policy statements, 2009–2012.* Washington, DC: NASW Press.

Ogden, O. (2008). Social work and the international humanitarian law: Rights, roles, and responsibilities. *Journal of Social Work Values and Ethics, 5*, 56–72. Retrieved from http://www.jswvearchives.com/content/view/90/65/

Pulla, V. (2013). Critical essay: Environmentalism and social work. *Rural Society, 22*, 263–268. doi:10.5172/rsj.2013.22.3.263

Reichert, E. (2011). *Social work and human rights: A foundation for policy and practice* (2nd ed.). New York, NY: Columbia University Press.

Reisch, M., & Andrews, J. L. (2002). *The road not taken: A history of radical social work in the U.S.* Philadelphia, PA: Brunner-Routledge.

Reisch, M., & Jani, J. S. (2012). The new politics of social work practice: Understanding context to promote change. *British Journal of Social Work, 42*, 1132–1150. doi:10.1093/bjsw/bcs072

Republic of South Africa. (1996). *Constitution of the Republic of South Africa.* Pretoria: Author.

Richmond, M. E. (1917). *Social diagnosis.* New York, NY: Russell Sage Foundation.

Sachs, J. D. (2009). *Common wealth: Economics for a crowded planet.* New York, NY: Penguin Group USA.

Schlosberg, D. (2007). *Defining environmental justice: Theories, movements, and nature.* New York, NY: Oxford University Press.

Schmitz, C. L., Matyók, T., Sloan, L. M., & James, C. (2012). The relationship between social work and environmental sustainability: Implications for interdisciplinary practice. *International Journal of Social Welfare, 21*, 278–286. doi:10.1111/j.1468-2397.2011.00855.x

Sexton, K., & Linder, S. H. (2010). The role of cumulative risk assessment in decisions about environmental justice. *International Journal of Environmental Research and Public Health, 7*, 4037–4049. doi:10.3390/ijerph7114037

Shaw, T. (2013). Is social work a green profession? An examination of environmental beliefs. *Journal of Social Work, 13*, 3–29. Retrieved at http://jsw.sagepub.com/content/13/1/3

Smith, V. (2014, July 22). Trash talk. *Baltimore City Paper.* Retrieved from http://www.citypaper.com/news/features/bcp-trash-talk-20140722,0,981899.story

Specht, H., & Courtenay, M. (1994). *Unfaithful angels: How social work abandoned its mission.* New York, NY: Free Press.

United States Environmental Protection Agency. (2014). *Statement on environmental justice.* Washington, DC: Author. Retrieved from http://www.epa.gov

Williams, T. (2015, 10 January). Garbage incinerators make comeback, kindling both garbage and debate. *The New York Times*. Retrieved from http://www.nytimes.com/2015/01/11/us/garbage-incinerators-make-comeback-kindling-both-garbage-and-debate.html

Worster, D. (2008). *A passion for nature: The life of John Muir*. New York, NY: Oxford University Press.

Zapf, M. K. (2009). *Social work and the environment: Understanding people and place*. Ontario: Canadian Scholars' Press.

Environmental Justice and Social Work Education: Social Workers' Professional Perspectives

Ande Nesmith & Nora Smyth

Environmental injustice is a growing human rights issue as climate change and environmental degradation rapidly increases. As a social justice problem, it is relevant to the social work profession, yet not integrated into our curricula. This study of 373 social work professionals found that environmental justice is a significant practice issue across broad client populations and that professionals felt unprepared to address it. Qualitative and quantitative data revealed high levels of client exposure to environmental hazards with little power to change it. Respondents reported dissatisfaction with their education to help them understand environmental issues. Moreover, they indicated that they would like to see environmental justice integrated into social work education and better-prepared graduates entering the profession. Implications for practice and education are discussed.

> The leadership we can provide with regards to environmental justice is powerful. We as social workers know how to effect change. We MUST continue to do so and to lead the charge in the realm of environmental justice for our clients and for ALL of our family members, be they two legged, four legged, winged, or leaf-covered. (Anonymous Study Participant)

Introduction

Climate change and environmental degradation may be the source of the most significant threat to human health, availability of life-sustaining resources, and basic human rights that the world has ever seen (United Nations Human Rights, 2014; World Health Organization, 2014). While there is extensive research on environmental hazards and injustices in impoverished nations, there is much less written about them in the context of wealthy countries. As such, in countries like the USA, social workers have not been trained to recognize or address environment-based risks and justice issues for their clients and communities.

It would be consistent with the history and mission of social work as a profession to be informed, engaged in the national and international discourse, and to contribute to the response to this threat. Yet, as a profession, social work has been slow to join the dialog at a policy, research, or practice level, leaving the work to a small group of committed social worker practitioners and scholars (Gray, Coates, & Hetherington, 2013). As a profession committed to social justice, social work is missing a critical opportunity to be actively engaged in fighting for equality and human rights.

The present study gathered data from one state of the USA to more closely examine the presence of environmental justice concerns and social worker reactions to them. This exploratory study observed the prevalence of environmental hazards and injustice in the context of a relatively wealthy state.

Definitions

Environmental justice is defined in many ways with no single agreed-upon definition. However, across the various iterations, there are two recurring themes: (1) the burden of environmental hazards or degradation is shared equally across all demographic groups or communities, and (2) there is equal inclusion in decision-making processes that result in environmentally related policies and actions. It is with this lens that environmental justice is discussed here.

Literature Review

Social work has its roots in human rights and social justice. According to the National Association of Social Workers, 'the adoption of human rights is a foundation principle upon which all of social work theory and applied knowledge rests' (, 2014). Environmental justice is as much a social justice issue as any other that social work has grappled with. In fact, fighting for environmental justice is not new to social work, though it may not have been framed in this language. Starting in the late 19th century, the pioneer social worker Jane Addams advocated heavily for environmental reform for the urban poor. Addams linked environmental justice and municipal reform in her fight for safe drinking water, a functional sewage system, garbage disposal, and clean, safe play spaces for children (Platt, 2000).

Relevance to the Poor and Marginalized in Wealthy Countries

Injustice is intrinsically tied to the impact of climate change and environmental degradation and exists in both developing and developed nations. It is imperative to recognize the role of disenfranchisement. Not only are those most affected found in the poorest countries of the world, but within those countries, it is the poor, women, and marginalized ethnic groups that suffer the gravest consequences—yet do not have a seat at the table where environmental policy decisions are made (Galizzi & Herklotz, 2013; Skwiot, 2008; United Nations Human Rights, 2014). This has perhaps made it tempting for wealthy nations to perceive environmental injustices as a problem belonging solely to developing countries.

Yet it is not only the vulnerable or marginalized in poor countries who suffer environmental injustices; this pattern is paralleled within rich countries as well. Environmental hazards are disproportionately experienced among low income and communities of color within rich nations. For example, a study in the UK examined the impact of cumulative and clustered environmental hazards such as flood zones, industrial sites, and active waste landfills and found that the poorest communities were most likely to be situated along these clustered risk sites (Fairburn, Butler, & Smith, 2009). In Canada, nitrogen dioxide air pollution was found to be heavily concentrated in poor neighborhoods while First Nations reserves suffer from contaminated drinking water (Buzzelli, 2008; Mascarenhas, 2007). Low income communities in the Netherlands experience more noise pollution, closer proximity to waste disposal sites, higher levels of air pollution, and limited access to green spaces relative to middle and upper income neighborhoods (Kruize & Bouwman, 2004). Similar patterns have been reported in Scotland, Sweden, France, Germany, and Australia (Scandrett, Dunion, & McBride, 2000; Higginbotham, Freeman, Connor, & Albrecht, 2010; Chaix et al., 2006; Kohlhuber, Mielck, Weiland, & Bolte, 2006; Laurian, 2008). The US offers no exception. Here too environmental hazards are disproportionately placed on those who have the least power and fewest resources to combat them, yielding strong correlations with race, economic class, and gender (McKibben, 2010). It is these communities that experience uncontrolled flooding or drought and unsafe drinking water, and house the industrial plants (McKibben, 2010; Tester, 2013).

Although the range of hazards in the USA is vast, according to the US Centers for Disease Control (2013) and the US Environmental Protection Agency (2012), the most common environmental hazards are air and water pollution. One known contamination source in the USA for both water and air is pesticide and herbicide use from farming, such as Atrazine which can cause cardiovascular damage, weight loss and cancer. Benzenes, which have significant short- and long-term health effects as well, are heavily used in manufacturing in the USA, ranking in the top 20 chemicals used by volume (US Centers for Disease Control and Prevention, 2013). Another source of air pollution affecting urban communities is automobile emissions. The contamination of water can be naturally occurring such as arsenic, or human-caused as with farming chemicals. About 10% of US citizens, primarily in rural areas, get their water from wells that are not treated or regulated. Other hazards that disproportionately affect some communities in the USA are hydraulic fracturing, 'fracking,' and natural disasters related to extreme weather such as hurricanes in southeastern and southern states along the Atlantic Ocean. Fracking is used extensively in the USA by using chemical mixes forced at high pressure into shale deposits to extract gases and is known to cause both earth quakes and leak toxic chemicals into water systems (Gosman, 2013; Shukman, 2013).

The US Midwestern state where this study occurred is vulnerable to all of the aforementioned hazards except hurricanes. The Midwest faces its own set of natural disasters, however, including flooding, tornados, and drought.

Hierarchy of Needs

Social workers engage in a broad spectrum of services, addressing a myriad of social and individual challenges. Many are trained at the master's level to conduct clinical work for individuals and families with mental health problems. However, as Maslow (1970) so famously framed it in his Hierarchy of Needs, basic needs, such as clean water, air, and adequate living spaces must be secured before one can address relationship concerns, confidence, or self-esteem. Clinical social workers in poor communities may have to devote more time tackling basic environmental needs and related justice issues than providing important therapeutic services. A limited scope of education can make it challenging to identify the sources of the problem and advocate on behalf of vulnerable clients.

Integrating the Natural Environment into Social Work Education

Education about environmental hazards and justice is critical to effecting change. An essential educational component that includes local examples is more apt to impact and motivate learners at the micro-level. When individuals understand a problem as personal or local, it feels more salient and urgent, and in turn leads more often to action—a process that has been documented with regard to climate change (Lorenzoni & Pidgeon, 2006; Spence, Poortinga, & Pidgeon, 2012).

There is a macro-level piece as well. How society views and supports a given perspective can influence action or inaction even in the presence of accurate information. A study of reactions to climate change in Norway found that while Norwegians are well educated with regard to their country's role in climate change, there is a process of 'socially organized denial' that perpetuates inaction. At the heart of this is a sense that to admit to knowing is in itself a political action (Norgaard, 2006).

While social work curricula are rigorous in confronting social, political, and economic inequalities, the field is still in the early stages of recognizing the expanding global environmental crisis and its disparate impact on the very people we assert to serve—the vulnerable, disenfranchised, and marginalized (Dominelli, 2012; Jones, 2013; Rogge, Davis, Maddox, & Jackson, 2005). To this end, this topic is driven by personal interests of select faculty and deans rather than by curricular standards. In the USA, social work curricula are governed by the Council on Social Work Education (CSWE), which establishes a competency-based educational framework. It is only at the time of this writing that the Educational Policy and Accreditation Standards (EPAS) of CSWE have recognized the importance of environmental justice in social work education. In the newly approved standards, environmental justice is now integrated into two of the eight required competencies, Competency 3, 'Advance

Human Rights and Social, Economic, and Environmental Justice' and as a bullet within Competency 4, 'Engage in Policy Practice.... critically analyze and promote policies that advance human rights and social, economic, and environmental justice' (CSWE, 2014).

Though still in the nascent stages, there is a palpable rising interest and discourse seen in academic publications on environmental justice and its role in social work education both nationally and internationally (Besthorn, 2013; Dewane, 2011; Gray et al., 2013; McKinnon, 2008; Rogge et al., 2005). As educators begin to recognize this topic as significant to social work curricula, it is still unclear how much social work practitioners view it as vital to their work. This is important because current practitioners are also the future employers of graduating students. Do these social workers and the organizations they represent see environmental justice as a valuable knowledge base for new social workers entering the field? If this kind of knowledge will make graduating social workers not only more effective but also more marketable, it could nudge schools of social work to go beyond the basic requirements of the new CSWE standards and fully integrate environmental justice throughout the curriculum.

To that end, this study sought to learn experiences and perspectives from current social workers on three areas: (1) the extent to which social work clients face environmentally related issues that are dealt with in practice, (2) how this affects practitioners' work and (3) the degree to which professionals want to see environmental justice content included in social work education.

Methods

Licensed social workers from a midwest state completed electronic surveys regarding their work-related experience with environmental justice issues, their opinions about how relevant and important these issues are for their clients, and their views on the relevance to social work education. The survey also included several open-ended questions.

Sampling

State-licensed social workers were selected using a randomly generated email list obtained through the state social work licensing board, which provides emails to the public upon request. Inactive and invalid emails were removed (about 10%). Another 100 emails were used for a pre-test. There remained 1,197 social workers with valid addresses to which the final draft was sent.

Measures

We pre-tested the survey on 100 randomly selected email addresses and received 21 responses, which were used to revise the questions. Revisions entailed minor adjustments to the order and phrasing of some items to make them clearer. The survey was distributed using Qualtrics survey software and covered three main topics: (1) the intersection of environmental justice and their social work practice; (2) specific examples of environmental hazards and environmental justice issues faced by their clients; and (3) the respondents' perspectives on integrating environmental justice into social work education.

The survey comprised a variety of question types. Some were Likert-type scaled items that were statements with answer choices ranging from 1 (strongly disagree) to 5 (strongly agree). Others were 'check all that apply' (in regard to types of hazards). The survey also included several open-ended questions. In this paper, we asked three open-ended questions pertaining to the above topics:

(1) *Clients*. Among those who indicated that their clients face environmental justice issues: What is the issue and how does it affect the clients?
(2) *Service*. How do these issues affect the way you provide services, if at all?
(3) *Education*. Would you like to see new social work graduates educated about environmental justice? If yes: What are the most important issues?

We examined a range of definitions for the terms 'environmental justice' and 'environmental hazards' at state, national, and international levels. A definition for each was created for this survey that avoided technical jargon while retaining the most pertinent and recurring content and language found in the various definitions. The survey provided respondents with the following definitions:

> *Environmental justice* occurs when all people equally experience high levels of environmental protection and no group or community is excluded from the environmental policy decision-making process, nor is affected by a disproportionate impact from environmental hazards.
>
> *Environmental hazards* refer to the presence of toxins or pollutants, natural disasters, lack of access to healthy foods, and other situations or activities that create an unsafe natural environment.

The survey listed ten environmental hazards and asked respondents to identify which, if any, were the most critical for their clients. The 10 hazards were the most prevalent issues in the state where the social workers were licensed. They were derived from a review of state reports, news articles, and other Internet sites that report on environmental issues. As not every hazard could be listed, respondents were also given the option to fill in other hazards not listed or to check a box for 'none' if they were not aware of their clients facing environmental hazards. While there are too many to itemize here all the hazards that were not named, some examples of omitted hazards are perfluorochemicals (human-made chemicals found in household products), radon gas, and more specific types of mining (e.g., taconite or sulfide).

Respondents

Of the 1,197 social workers who were sent the survey, 373 (31%) social workers completed it. To retain anonymity, the survey did not ask respondents to name their city. However, the sampling process included all counties in the state; the qualitative responses revealed that respondents represented rural, suburban, and urban regions of the state. All respondents indicated that they were currently engaged in professional social work practice.

The majority (57%) of respondents had an MSW, while 28% had a BSW as their highest degree. The remaining either had a doctoral level education or did not answer the question. Fifty-four percent held a generalist practice license at either the bachelor's or master's level. Six percent were graduate-level licensed independent social workers, and 39% held a clinical license (LICSW).

Respondents reported an average of 15.9 years of experience in social work practice. While this was not heavily skewed (median = 15 years), there was a wide range of experience, from less than a year to 45 years.

Analysis

This is a cross-sectional exploratory study examining the views of practicing social workers, rather than an explanatory study seeking to find a cause–effect relationship. Therefore, the quantitative analyses were primarily descriptive at the univariate and bivariate levels. Bivariate analyses included *t*-tests and chi-square.

Qualitative Analysis

One or more of the open-ended questions were answered by 249 social workers. While most were brief, some provided great detail. The qualitative analysis was performed by the primary researcher and a graduate research assistant. The qualitative responses presented a substantial range in terms of breadth and detail. However, because these were not in-person interviews, most of the qualitative answers did not have great depth nor a personal narrative. The survey mode and anonymous nature of it did not permit follow-up questions. For open-ended responses listing environmental hazards or environmental justice issues, we grouped them categorically by types of hazards or issues presented, cross-checking our findings with each other for consistency.

Other responses were stories about how issues unfolded and how the respondent handled them. These were analyzed using content analysis, which is an appropriate method for open-ended survey questions (Hsieh & Shannon, 2005). This is a systematic approach to identifying thematic categories by creating codes that use specific words or phrases as the 'data.' It is important to retain the context to avoid over-simplifying coded themes (Cho & Lee, 2014). Therefore, we report most or all of the passages surrounding the coded phrases in our findings.

We began using an emergent coding process. Each researcher first independently reviewed the open responses and developed a coding scheme. We arrived at a list of elements that suggested a theme and identified all the terms and phrases that fit with it. A cross-check of the codes and terms that defined them were then consolidated until we arrived at a list that was consistent without redundancies (Elo & Kyngäs, 2008). We began by coding themes pertaining to broad topics such as physical hazards or mental health risks. As we dug deeper, we observed examples of underlying feelings and experiences such as a sense of powerlessness or efforts of advocacy. This was an iterative process that included re-reading the transcripts after we developed the codes, to identify and confirm all instances for each theme. Using a constant comparison approach, we compared findings and jointly determined whether the evidence confirmed or disconfirmed the theme, or was unrelated (Miles, Huberman, & Saldana, 2014).

Human Subjects Ethics Review

This study was reviewed by the primary author's university institutional review board and was approved prior to data collection. The survey software removed identifying information from respondents so that it was not possible to link any response to any given email address or individual.

Findings

The presentation of the qualitative and quantitative findings are integrated to reveal both the numeric responses and quotes that exemplify how they were experienced by the social workers. The findings begin by presenting the nature and extent of the problem: perspectives on the salience of environmental justice to the profession, the range and importance of environmental hazards, and issues of power imbalance and social justice. Then, we examine suggestions for action and solutions: perspectives on advocacy, change, and better education for new social workers.

Relevance to Professional Social Work Practice

Respondents were asked to consider the relevance of environmental justice issues to social work. They rated their agreement with statements on a scale of 1 (strongly disagree) to 5 (strongly agree). Table 1 presents the mean scores for these items.

On the whole, they tended to view environmental justice as important to the profession (mean = 3.93). When asked about their personal experience with environmental justice, respondents tended to disagree (mean = 2.29) with statements indicating that environmental justice is not relevant to their work or their clients or that they do not think about it when working with clients

Table 1 Relevance to professional practice.

	Mean
Environmental justice should play a significant role in social work practice	3.93
Environmental justice is primarily a problem in poor countries rather than in the U.S	1.84
Environmental justice is not relevant to my current social work practice or clients	2.29
Even if environmental justice is relevant or important to my clients, it is not my role to address those issues	2.10
I do not think about the role of environmental justice when working with clients	2.66
Would like to focus more on environmental justice issues, but my employer does not support it	2.62

(mean = 2.66). As a large share of academic writing on environmental justice cites examples from countries that are poorer than the USA, this can be a lens through which social workers perceive the topic. We asked to what extent they agreed that environmental justice was primarily an international issue beyond the USA and found that nearly all disagreed or strongly disagreed (mean = 1.84). However, there was some indication that their perspectives were not representative of all social workers. One respondent wrote,

> I think so many newer social workers and Americans for that matter believe that these are 'other countries' issues! Many people don't even realize the impact on our communities here.

Finally, we explored whether employers stood in the way of social workers focusing on this topic and learned that this was of minimal concern (mean = 2.62).

Critical Environmental Hazards for Clients

Respondents were asked to identify the environmental hazards that were most critical to their clients, if any (Figure 1). Only 8% reported that their clients did not face any such hazards.

Limited access to healthy food (food deserts) was the most common response (69%). This was followed by lack of safe play spaces for children (51%), lead poisoning, air pollution, extreme weather, and farming pesticides. Although safe drinking water was less often ranked a critical hazard, it was one of the most-discussed issues in the open-ended questions. Perhaps, this is because it is so vital to survival, as one respondent explained, 'Access to safe water: It's such a basic human need that without protection, everyone is vulnerable.'

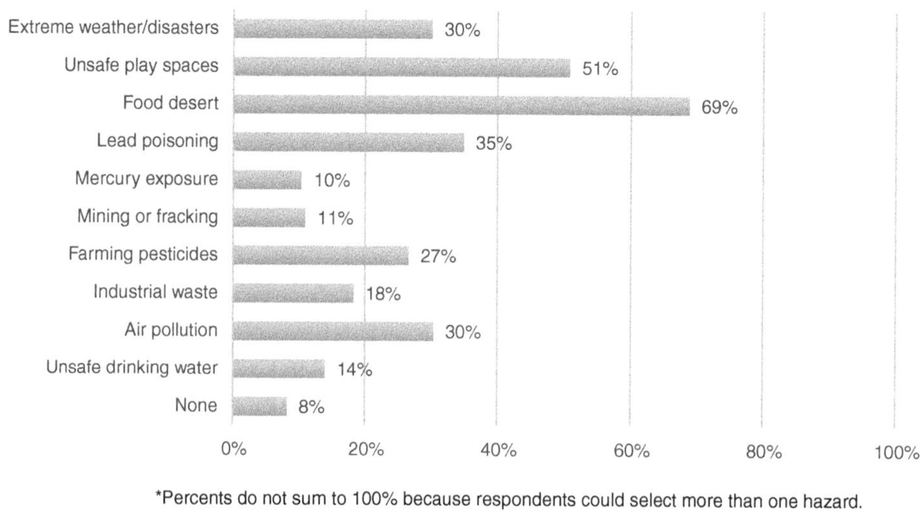

Figure 1 Most critical environmental hazard for clients ($n = 373$)*.

There were a number of additional comments made in regarding to safe drinking water. One of the most common concerns was unsafe levels of arsenic, typically found in rural areas with well water. The respondents shared stories in which tenants, especially those in mobile home parks, were given a choice to either drink the poisonous water or purchase bottled water that they could not afford.

> I had a client that lived in a mobile home park. There was no safe drinking water, except for one water spigot in the whole park. Everyone had to take their gallon jugs there to get water. It was especially difficult for those who were elderly or disabled to get to the water.

The concern regarding play spaces was tied both to air pollution or absence of them, noting that the poorer clients living in some of the most polluted areas

> [My clients] tend to be located near industrial areas with higher levels of pollution... the neighborhoods they can move into do not have playgrounds or the air is too polluted to be outside... this can also contribute to obesity in children, as parents keep their kids indoors where they have limited ability to get exercise.

Environmental Injustice

The survey asked if the respondents believed that their clients faced environmental injustices. Seventy-one percent of respondents reported that they had a client facing environmental justice issues.

Respondents provided examples of injustice, the lack of client control over the situations, and sometimes their own feelings of helplessness. Coded terms that revealed injustice or lack of power included words like 'helpless,' 'hopeless,' or 'powerless' in reference to the client experiences. For their own feelings, social workers used language to describe a lack of time, resources, or support. Table 2 outlines some examples of how this language was used to support the themes. The coded words or phrases used to identify the theme are in italics.

It was evident from the extent and range of responses that the power imbalance and impotence to effect change were felt acutely by both clients and social workers.

Table 2 Injustice and powerlessness.

Theme	Examples of language supporting theme
Client feelings of lack of control over situation	There is a lack of a sense of safety in their environments and feeling *powerless* to change risk factors They feel *helpless* as they do n't believe they have any recourse or way of obtaining a satisfactory outcome The people most affected do not have a voice in the process. This can affect their quality of life and sense of well-being. It often leads to *hopelessness*
Social worker feeling lack of power	I spend *more time using Band-Aids* and crisis management I *do not have a lot of answers* or responses to their concerns other than to provide some on-going reassurance *Time is limited* and *support* to really be a social worker is just *not there.* Honestly, at times there are *no resources* to be offered which could eliminate or minimize the environmental injustice

Hierarchy of Needs

Many respondents saw a direct relationship between environmental hazards and their ability to help clients with socio-emotional needs. Several specifically mentioned Maslow's Hierarchy of Needs as they discussed the challenges of helping clients with mental health issues when basic environmental needs were threatened. Table 3 presents examples of language that supported the theme of the hierarchy of needs.

Most of the social workers who raised the concerns about basic needs also expressed frustration that they wanted to be able to work with clients on mental health issues but spent much of their time and energy getting them safe drinking water, healthy food, or residences away from other toxins.

Need for Action and Advocacy

Participants provided a plethora of ideas of what they envisioned could improve client situations or shared what they were already doing to address environmental injustice (Table 4). These fell into four main categories. The first is community organizing. Social workers are taught how to work with and organize large client systems. These respondents drew on that knowledge to suggest ways to help communities organize at the grass roots level. The second category is training. Some of the respondents saw a gap in their knowledge and elected to get the information and training they needed to better support their clients on environmental injustices or on environmental hazards specific to their region. Third was drawing on the advocacy that is central to the profession and addresses the social justice mission of social work.

Finally, the fourth category includes other actions that did not fit in the first three but were clear statements of intentional behaviors to achieve a positive change.

Table 3 Inadequacy of basic needs interferes with mental health.

Theme	Examples of language supporting theme
Basic needs	I have to spend more time on case management getting them services rather than providing therapy. They can't concentrate on improving mental health until *basic needs* are met
	Basic needs need to be met before people feel safe and can work on emotional needs above and beyond basic needs—diet and nutrition, and correlates with exacerbated mental health issues
Hierarchy of needs	I liken it to *Maslow's* hierarchy of need theory
	Maslow's hierarchy of needs—if a client does not have safe housing, they are not going to be concerned about working on other goals related to their health care or mental health stability
Environmental issues impede health	*Limited access* to healthy food and clean water makes it difficult to motivate clients to make *healthy lifestyle choices*
	There are situations in which people find themselves living in *environmental conditions* which have a direct impact on their ability to live *happy, successful* lives
	Struggling with unsafe or unhealthy surroundings and inadequate or *unhealthy food and water* takes energy away from the client's ability to move forward with *life goals*

Table 4 Taking action.

Theme	Examples of language supporting theme
Community organizing	The importance of *non-institutional tactics* most conducive to *grassroots* power such as *mass meetings*, demonstrations, disobedience, and picket lines As *communities* and agencies, we should be *supporting local* farmers markets and working with them to provide affordable, healthy food for everyone
Training	I am *seeking training* to increase my capacity to bring the issues into the consulting room with more regularity and impact I was not trained to focus on these issues and am *seeking the training* to do so now
Advocacy	... the causes of the problem continue to expand rapidly inevitably rousing social workers into the role of *advocacy* in relation to environmental justice issues. We constantly need education, *advocacy* and protection for all, especially the more vulnerable The *leadership* we can provide with regards to environmental justice is powerful. We as social workers know how to *effect change.* We MUST continue to do so and to *lead the charge* in the realm of environmental justice for our clients...
Other Actions	We must *validate* reality of situation, validate client's assessment, feelings, *teach* problem solving, access resources, *instill* hope, *coach* assertiveness, and *communicate* with other providers Anything that can be done to *raise awareness* and *positive attention* to taking better care of our planet will result in better outcomes and can actually be a economically helpful tool at the same time if investments are done

Changing Education to Better Prepare New Social Workers

We asked respondents to respond to statements about environmental justice in their own social work education and social work education in general (Table 5). Respondents disagreed that content on environmental justice was a significant part of their education and indicated that they did not receive adequate training to identify environmental hazards or attend to environmental injustice (means 1.58 to 2.10). They agreed that the profession would benefit from more content in this area (mean = 3.95).

Respondents were asked as active social work professionals, to think about the kind of education and training they might want to see in new employees at their place of

Table 5 Environmental justice and social work education.

	Mean
Individual educational experience	
Content on environmental justice was a significant part of my social work educational program	1.58
I was adequately trained in how to identify important environmental hazards for clients	2.10
I was adequately trained in how to address environmental justice issues in social work practice	2.10
Social work education in general	
The profession would benefit from more content on environmental justice in social work education	3.95
In general, social workers are adequately educated in environmental justice issues	2.11

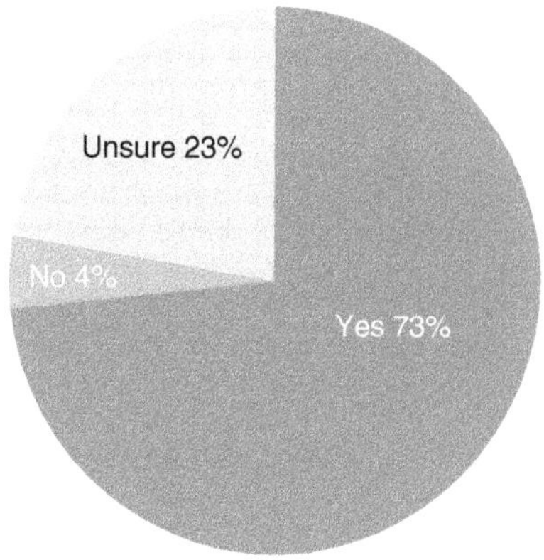

Figure 2 Should new social work graduates be educated about environmental justice?

work. Respondents were asked if they were to hire a new social work graduate if they would like to see that person educated about environmental justice. Nearly three-quarters (73%) responded 'yes,' with 23% unsure and 4% indicating that this was not important (Figure 2).

We investigated whether the level of education or years of experience were associated with respondents' views on whether environmental justice should be integrated into social work education. Chi-squared tests revealed no significant difference based on the respondent's degree (BSW, MSW, Doctorate) and *t*-tests found no significant differences based on how many years they had been a social worker.

Respondents who had clients facing environmental justice issues were significantly ($p < 0.000$) more likely than others to want future social work graduates educated in environmental justice. Interestingly, even among those who reported that they did not have a client with an environmental justice issue, 54% reported that they would like to see social workers educated on the topic.

Participants were given an opportunity to share ideas about incorporating environmental justice into social work education. The most common responses were to link it to systems theory, to suggest that this is relevant to all people, or to provide specific ideas about how education can improve practice.

Limitations

The most significant limitation to this study is that it was only conducted in one state in the USA. While neighboring states likely experience similar environmental hazards, this sample does not represent the broad range of hazards that exist across the USA, especially the coastal states. Moreover, the results cannot be generalized beyond the US Justice issues, in particular manifest quite differently in different nations. Surveys are notorious for low response rates, which introduce sampling bias. It is likely that the findings here over-represent social workers who are concerned about environmental issues. Those who do not view the natural environment as relevant, or who work with clients who are not very affected by it, may be less inclined to respond. This study is cross-sectional and cannot speak to causal inference between variables. Finally, because the survey was anonymous, we were not able to follow-up with clarifying questions for any of the open-ended items.

Discussion

On the whole, the participants in this study believed that addressing environmental injustices was relevant to the profession, that they received inadequate training on it, and that it should be integrated more intentionally into social work education. Despite being in an economically stable state that does not have some of the more dramatic natural disasters, nor some of the abject poverty seen in other states and other countries, 92% of the respondents were working with clients who confront environmental hazards. The messages shared by the participants indicated that clients felt powerless over their environmental situations, a feeling also reflected amongst the social workers in the face of limited resources or power. Perhaps, the strongest and most prevalent message was that environmental challenges, whether unsafe water, toxins in the air, or limited access to healthy food, impeded the therapeutic process for those with mental health concerns.

Yet, importantly, there was also a strong message of hope and action. Although many social workers felt pressed for time and resources, it was encouraging that their employers did not discourage them from addressing these issues. They sought ways to organize vulnerable communities, to take the initiative to educate themselves on related topics that they never were trained in, and advocating on their clients' behalf when possible. Equally critical, they wanted to see future social workers, ones whom they might employ, to be competent in this area.

Implications for Practice

The source of environmental injustice in wealthy countries is multi-faceted. It is tied to poverty, lack of mobility, lack of power, and racism as evidenced by countless studies. Poverty limits mobility, forcing residents to live in low-cost areas that are also known to have hazards such as flood risks or contaminated air and water (Hunter, 2005). It is also the case that poor communities are selected as industrial or hazardous waste sites expressly due to their limited political or social power. When this occurs, moving away can be an expensive and unattainable solution, and may only lead them to different hazardous locations (Hunter, 2005).

Our findings indicate that even in a first-world country like the USA and in fairly wealthy regions, environmental injustice can be prevalent while education to prepare for it is inconsistent and insufficient at best. Working toward environmental justice can draw in part on skills that social workers already have in the USA, such as advocating for clients and helping them acquire resources to escape poverty. Those are valuable micro-level goals. At the macro-level, social workers may need to use their community organizing skills to transition communities into more powerful positions in the decision-making processes. Social work education needs national standards that exist not only in name, but that require enough knowledge about environmental hazards, their implications for client health, and the political structures driving environmental decisions that social workers can spark genuine change at both micro- and macro-levels.

The new EPAS standards will likely result in mention of environmental justice and the natural environment as another system layer in human behavior courses. This will be an important start, but it will not be adequate by itself. Social work students need real-life examples of advocacy and community organizing to help their clients move beyond addressing their basic needs to meet higher-order socio-emotional goals. Students need field placements that focus on environmental conditions. When they take courses on large client systems, they can be exposed to local examples of community organizing around environmental injustices, both urban and rural. As some have claimed, to gain both legitimacy and competence, our students must acquire an understanding of the science behind climate change, the science of environmental toxins and pollutants, and the complex relationships different populations have with the environment (Dominelli, 2011; Tester, 2013).

Implications for Research

So little research has been conducted on the role of social workers in environmental justice that it goes without saying that more is needed. This study was exploratory and conducted in a limited geographic location. It is important first to replicate the study in other states in the USA that are known to experience a different variation of environmental hazards, such as the east coast and south, which suffers from increasing hurricanes and other severe storms. Studying the west coast of the USA could shed light on the impact of longstanding droughts and injustices on water availability or the impact of wildfires. Alaska has its own brand of environmental issues with arctic drilling and risks of oil spills. While one can always legitimately argue for the need to research more in impoverished countries, part of the value of this study is highlighting the fact that environmental hazards and injustices are everywhere, and are not only in the domain of developing nations.

As social work education embarks on required competency in environmental justice, it will be important to conduct research to evaluate the extent to which it is integrated into the curriculum and how prepared graduates feel to address these concerns as they enter the field. Environmental hazards and the impact of climate change represent a constantly evolving knowledge base as science attempts to keep pace with new information. Equally important is that environmental policies both nationally and globally are continually changing. Social work education will need to stay apprised of the most current knowledge and ensure that its students are adequately educated.

Conclusion

Given the current climate change trajectories, these issues promise to grow exponentially in the coming years (McKibben, 2010). There is little excuse to continue to ignore environmental justice professionally in a time when climate change and its

Table 6 Integrating environmental justice in social work education.

Theme	Examples of language supporting theme
Systems theory	This is certainly a *systems-based issue* which should be addressed at least in a basic way in all social work educational programs! It is vitally important that we as humankind recognize our interdependency with all of life. In our work as social workers, this interdependency should be understood as fundamental to any work we do. It is indeed a 'systems framework' approach!
General need for education	*Education* is the key *across the board* to professionals as well as clients Think it would be a great addition to a *social work grad program*, and probably warrants its own specific track
Specific curriculum goals	. . . *learn how* environmental hazards should be included in the initial assessment, *know what* are their client's risk factors and *be able to* incorporate environmental justice issues into diagnosis treatment and case plans Students *should learn* (1) Basic knowledge; (2) conservation of resources; (3) energy; (4) grassroots organizing to get those who are affected disproportionately involved and educated about issues It is critical that social workers *receive education on how to* incorporate environmental justice and eco-psycho-therapeutic principles into assessment and treatment planning with clients. Whether or not clients name these specific issues, they are impacting their lives

repercussions are recognized by global science and political communities as genuine and intensifying threats to humanity and all life on Earth. Rather than waiting until there is no choice but to join the ranks, the profession has an opportunity to be actively involved in confronting this problem and raising awareness.

Historically social workers have seized the opportunity to be both the voice for the voiceless and to organize vulnerable communities to have their own voices heard. As seen in the open responses of this study, the energy and drive to create change is already present among current social workers. Now is the time to recognize the implicit connections between the environment and human rights, so that we can envision not only a world-wide culture of environmental sustainability, but also one that is fair and just for all (Table 6).

Disclosure statement

No potential conflict of interest was reported by the authors.

References

Besthorn, F. (2013). Radical equalitaraian ecological justice. In M. Gray, J. Coates, & T. Hetherington (Eds.), *Environmental social work* (pp. 31–35). New York: Routledge.

Buzzelli, M. (2008). Environmental justice in Canada – It matters where you live. In *Canadian Policy Research Networks Research Report*. Ottawa, Ontario: CPRN.

Chaix, B., Gustafsson, S., Jerrett, M., Kristersson, H., Lithman, T., Boalt, A., & Merlo, J. (2006). Children's exposure to nitrogen dioxide in Sweden: Investigating environmental injustice in an egalitarian country. *Journal Of Epidemiology & Community Health, 60*, 234–241. doi:10.1136/jech.2005.038190

Cho, J. Y., & Lee, E-H. (2014). Reducing confusion about grounded theory and qualitative content analysis: Similarities and differences. *The Qualitative Report, 19*(64), 1–20.

CSWE. (2014). Draft 3 of the 2015 Educational Policy and Accreditation Standards (EPAS). Retrieved on June 6, 2015 from http://www.cswe.org/Accreditation/EPASRevision.aspx

Dewane, C. J. (2011). Environmentalism & social work: The ultimate social justice issue. *Social Work Today, 11*, 20–22.

Dominelli, L. (2011). Climate change: Social workers' roles and contributions to policy debates and interventions. *International Journal of Social Welfare, 20*, 430–438. doi:10.1111/j.1468-2397.2011.00795.x

Dominelli, L. (2012). *Green social work: From environmental crises to environmental justice.* Cambridge, MA: Polity Press.

Elo, S., & Kyngäs, H. (2008). The qualitative content analysis process. *Journal of Advanced Nursing, 62*, 107–115. doi:10.1111/j.1365-2648.2007.04569.x

Fairburn, J., Butler, B., & Smith, G. (2009). Environmental justice in South Yorkshire: locating social deprivation and poor environments using multiple indicators. *Local Environment, 14*(2), 139–154.

Galizzi, P., & Herklotz, A. (2013). Missing in action: Gender in international environmental law. In C. Blerta, I. Dankclman, & J. Stern (Eds.), *Powerful synergies: Gender equality, economic development, and environmental sustainability* (pp. 29–41). United Nations Development Programme. Retrieved from http://www.undp.org/content/undp/en/home/librarypage/womens-empowerment/powerful-synergies.html

Gosman, S. (2013). Reflecting risk: chemical disclosure and hydraulic fracturing. *Georgia Law Review, 48*, 83–144.

Gray, M., Coates, J., & Hetherington, T. (2013). Overview of the last ten years and typology of environmental social work. In M. Gray, J. Coates, & T. Hetherington (Eds.), *Environmental social work* (pp. 1–28). New York, NY: Routledge.

Higginbotham, N., Freeman, S., Connor, L., & Albrecht, G. (2010). Environmental injustice and air pollution in coal affected communities, Hunter Valley, Australia. *Health & Place, 16*, 259–266. 10.1016/j.healthplace.2009.10.007

Hsieh, H., & Shannon, S. E. (2005). Three approaches to qualitative content analysis. *Qualitative Health Research, 15*, 1277–1288. doi:10.1177/1049732305276687

Hunter, L. M. (2005). Migration and environmental hazards. *Population and Environment, 26*, 273–302. doi:10.1007/s11111-005-3343-x

Jones, P. (2013). Transforming the curriculum: Social work education and ecological consciousness. Overview of the last ten years and typology of environmental social work. In M. Gray, J. Coates, & T. Hetherington (Eds.), *Environmental social work* (pp. 213–230). New York, NY: Routledge.

Kohlhuber, M., Mielck, A., Weiland, S. K., & Bolte, G. (2006). Social inequality in perceived environmental exposures in relation to housing conditions in Germany. *Environmental Research, 101*, 246–255. 10.1016/j.envres.2005.09.008

Kruize, H., & Bouwman, A. A. (2004). Environmental (in)equity in the Netherlands. A case study in the distribution of environmental quality in the Rijnmond region. In *National Institute for Public Health and the Environment.* Netherlands: Rijnmond.

Laurian, L. (2008). Environmental injustice in France. *Journal of Environmental Planning and Management, 51*, 55–79. doi:10.1080/09640560701712267

Lorenzoni, I., & Pidgeon, N. F. (2006). Public views on climate change: European and USA perspectives. *Climatic Change, 77*, 73–95. 10.1007/s10584-006-9072-z

Mascarenhas, M. (2007). Where the waters divide: First nations, tainted water and environmental justice in Canada. *Local Environment, 12*, 565–577. 10.1080/13549830701657265

Maslow, A. (1970). *Motivation and personality* (2nd ed.). New York, NY: Harper & Row.

McKibben, B. (2010). *Eaarth: Making a life on a tough new planet.* New York, NY: St. Martin's Griffin.

McKinnon, J. (2008). Exploring the nexus between social Work and the environment. *Australian Social Work, 61*, 256–268. doi:10.1080/03124070802178275

Miles, M. B., Huberman, M. A., & Saldana, J. (2014). *Qualitative data analysis: A methods sourcebook.* Los Angeles: Sage.

NASW (2014). International policy on human rights. Retrieved from https://www.socialworkers.org/pressroom/events/911/humanrights.asp

Norgaard, K. M. (2006). "We don't really want to know": Environmental justice and socially organized denial of global warming in Norway. *Organization & Environment, 19*, 347–370. doi:10.1177/1086026606292571

Platt, H. L. (2000). Jane Addams and the ward boss revisited: Class, politics, and public health in Chicago, 1890–1930. *Environmental History, 5*, 194–222. ISSN 1084-545310.2307/3985635

Rogge, M. E., Davis, K., Maddox, D., & Jackson, M. (2005). Leveraging environmental, social, and economic justice at chattanooga creek. *Journal of Community Practice, 13*, 33–53. doi:10.1300/J125v13n03_03

Shukman, D. (2013). What is fracking and why is it controversial? BBC UK. Retrieved from http://www.bbc.com/news/uk-14432401

Skwiot, R. (2008). Green dream: Environmental justice is emerging from the shadows. *Social Impact,* Fall, 23–27.

Spence, A., Poortinga, W., & Pidgeon, N. (2012). The psychological distance of climate change. *Risk Analysis, 32*, 957–972. doi:10.1111/j.1539-6924.2011.01695.x

Tester, F. (2013). Climate change as a human rights issue. Overview of the last ten years and typology of environmental social work. In M. Gray, J. Coates, & T. Hetherington (Eds.), *Environmental social work* (pp. 102–118). New York, NY: Routledge.

U.S. Centers for Disease Control and Prevention. (2013). Retrieved from http://ephtracking.cdc.gov/showRiskLandingSolution.action *Environments.*

U.S. Environmental Protection Agency. (2012). *Water science: Drinking and water systems?* Washington, DC: U.S. Environmental Protection Agency. Retrieved from http://www2.epa.gov/science-and-technology/water-science

U.S. Environmental Protection Agency. (2014). *What is environmental justice?* Washington, DC: U.S Environmental Protection Agency. Retrieved from http://www.epa.gov/environmentaljustice/

United Nations Human Rights. (2014). *Human rights and climate change.* Office of High Commissioner for Human Rights. Retrieved from http://www.ohchr.org/EN/Issues/HRAndClimateChange/Pages/HRClimateChangeIndex.aspx

World Health Organization. (2014). *Climate Change and Health.* Retrieved from: http://www.who.int/mediacentre/factsheets/fs266/en/

Changing Gears: Shifting to an Environmental Perspective in Social Work Education

Mel Gray & John Coates

This paper seeks to add to the growing literature on environmental social work education which suggests the need for a fundamental rethinking of the humanistic values and theories informing social work to embrace concerns relating to environmental degradation and climate change. For the most part, social work's interest in the environment to date relates to human needs. Of most concern here is the over-representation of people in poverty and subsistence among those impacted by deforestation and climate injustice. However, even here the emphasis is on the human experience of environmental and climate change when this is an outcome of human actions and structural inequalities. The paper begins with an overview of the theoretical terrain of environmental thought before examining issues in relation to perspective transformation and the implications for under- and post-graduate curriculum development.

The International Declaration of Human Rights enshrines key values that undergird social work analyses of social policies and their need to mitigate the worst effects of global economic growth and development. In supporting free trade and globalization, national governments and international development organizations have colluded to promote—directly and indirectly—an extractive economy and the destruction of major ecosystems, not to mention the exacerbation of gross inequalities. The literature discusses that environmental impacts are directly related to poverty and detrimentally affect those very populations that social workers seek to help, especially the poor. For example, Pandey (1998) discussed the relationship between women, the environment and sustainability in Nepal, where rural women and children were overwhelmingly threatened by environmental degradation and deforestation. Case examples such as this underscore the importance of an environmental perspective for social workers seeking to improve the plight of individuals, families and communities who not only bear the brunt of pervasive inequality and social injustice, but also are doubly disadvantaged by adverse environmental events.

The environmental movement has not only exposed how human action and exploitation contribute to environmental change and structural inequality, but also provided viable alternative values of conservation, degrowth, diversity, sustainability, and restoration modelled on the vision of a thriving planet with flourishing inhabitants (Berry, 1988, 1999; Capra, 1982; Hannis, 2015; Korten, 2009; Spretnak, 1997). These values offer social work a framework for advancement towards equality and inclusion that point to transformative social work education and practice (Coates, 2003, 2005; Dominelli, 2012, 2013; Ife, 2013). This paper concerns itself with a vision of social work as 'a political and moral enterprise' (Clark & Asquith, 1985, p. 2) concerned with justice, change and social transformation, and how environmental values might shape a new curriculum (Coates & Gray, 2012; Gray & Coates, 2012). In this way, it seeks to add to the growing literature on environmental social work education (Besthorn, 2003; Besthorn & Canda, 2002; Hayward, Miller, & Shaw, 2013; Hoff, 1992; Jones, 2008, 2010, 2013, 2014; Kahn & Scher, 2002; Rogge, 1993; Schmitz, Matyok, James, & Sloan, 2013).

An Opening Note on Terminology

There is a confusing array of terminology in the environmental literature, which, overall, is not, by definition, holistic. As Ife (2013) correctly points out, there are two distinct strands of environmental thought. The strand he describes as 'environmental' is based on linear cause–effect thinking and, by definition, is not holistic. He distinguishes this 'environmental perspective', which he says, seeks change within the existing modern social and economic order, from a more radical 'Green perspective' pushing for a fundamentally transformed society. Together, Ife's (2013) 'environmental' and 'Green' approaches comprise what he refers to as an 'ecological perspective' that is very different from its early iterations in social work, where the terms 'ecological' and 'ecosystems' were used as an extension of the 'person-in-environment perspective' (e.g. Germain, 1981, 1991; Germain & Gitterman, 1980; Meyer, 1983). These early writers focused on the person in the *social* environment drawing on Bronfenbrenner's (1979) ecology of human development, and their so-called 'ecological' approaches had little concern for what environmentalists see as the *physical* environment. Hence these social work scholars, among others, appropriated and applied the ecological model without truly understanding the full spectrum of interdependence, diversity and community that goes with a 'holistic ecological perspective' (see Besthorn, 1997; Coates, 2003; Ife, 2013; Zapf, 2009).

Given the confusion surrounding the use of the term 'ecological' in social work, and the fact that the term 'environmental perspective' does not necessarily refer to holism and interdependence, what might a more accurate term be? Possibly 'deep ecological' or 'dark green', which, as Besthorn (2012) describes, arose in opposition to the linear environmental approach, is more accurate, but this, too, like 'Green', has related implications. The term 'Green' has come to be associated with Left-leaning political parties on the back of a social-environmental movement that includes diverse social and political groupings, organizations and actors from the grassroots to high-level international players. Green has also been used as a marketing tag to suggest that things that are anything but 'environmentally friendly', such as companies manufacturing doors and paper, and draws on the emotive notion that 'green' means clean, good and wholesome, such as green living and thinking. One rather complex option is an 'enriched environmental, holistic ecological' perspective but, given ecology's association with the 'science of ecology'—as in 'is it ecologically sound?'— in this paper, we have opted for a more inclusive 'holistic environmental' perspective, which is, no doubt, also debatable.

The Theoretical Terrain of Environmental Thought

Despite the confusion surrounding terminology already described, the alternative theoretical framework—assumptions, values and beliefs—that informs a holistic environmental perspective builds on a long line of scholars from several disciplines (e.g. Berry, 1988, 1999; Capra, 1982; Coates, 2003; Hart, 2004; Spretnak, 1997; Trainer, 1985, 2010), who discuss its core features of holism, diversity, sustainability, equilibrium and interdependence. These are quite different from modern society's priorities that include economic growth, determinism, anthropocentrism, consumption and individualism. Drawing attention to, and fully articulating the nature of, this holistic environmental perspective have been significant contributions of environmental thought for social work (see, for example, Besthorn, 2013; Dylan, 2013). A holistic environmental perspective points to the limitations of linear thinking in modern science, and society's unquestioned faith that science and technology can provide all the answers to the environmental problems that we are currently facing. Further, it points to the unsustainability of the existing socio-economic order and the consequences it has wrought for the environment. Most importantly, it reflects an alternative foundation of values and assumptions, which requires a deep exploration of social work's humanistic values and theoretical foundations.

The challenge with any form of transformational practice and education is to critique, yet build on, what has gone before. Up until now, much critical literature has served to polarise thought by rejecting prior analyses, creating dichotomies and generating unnecessary conflict, rather than creating a more progressive and integrative analysis and practice. The biggest and most important contribution of the environmental discourse for social work has been the critical exploration of social work's role in modern society and the humanistic assumptions, beliefs and values that support this role. Such a critique opens the door to the profession seeing the limitations and boundaries of modernity (dualism, uncaring domination, ruthless exploitation, and reductionism) that have constrained social work thinking and action to the primary activities of helping *people* 'to fit in and adjust to the demands and expectations of modern life' (Coates, 2003, p. 38) rather than to change them.

Holistic environmental discourse supports the exploration of an alternative and inclusive set of values, beliefs and assumptions that include interdependence, community and diversity (see Berry, 1988; Besthorn, 2013; Coates, 2003; Mary, 2008; Spretnak, 1997). These values are seen as important to living in right relationship to Earth so as to promote the integrity and stability of the natural world on which human development and survival depends (Brown & Garver, 2009). Human dependence on the physical environment sits behind social work's goal to help create sustainable environmental conditions for the flourishing of the human—and natural—world (Hannis, 2015). Within a holistic environmental perspective, individual well-being is understood as embedded in community, that is, in 'the well-being of my neighbour, of all of the planet's people and of the Earth itself' (Coates, 2003, p. 156). Further, it requires a critical perspective on the impact of human actions on environmental change and structural inequality. This broader perspective distances social work from the constraints of economic and materialistic determinism and places it in the struggle for social transformation within a new foundation of values beyond concerns for human interests (see Coates, 2003; Gray & Coates, 2012; Jones, 2013). Thus, perspective transformation becomes a central component of social work education as it lies at the heart of a transition from a modernist, human-centred to an enriched holistic environmental perspective.

Perspective Transformation

Mezirow's (1991, 1997, 2012) work on perspective transformation outlines several steps that educators can take to help create conditions that support a shift in perspective. These steps include exposure to, and critique of, the dominant perspective and the presentation of an alternative one. Any curriculum change towards a holistic environmental approach would benefit by recognising and accommodating this process (Coates, 1994). Perspective transformation is, most frequently, slow and gradual, because people do not easily let go of deeply entrenched values and assumptions. A new perspective throws us out of our comfort zone and requires changes in lifestyle and behaviour that many are reluctant to make. However, tides of change in collective thinking can support a shift in assumptions, values and actions. We think environmental awareness is one such tide that will eventually make it impossible for social work to avoid making fundamental changes in education and practice.

To be effective, the introduction of environmental issues into the social work curriculum would require support within a school of social work though, in the USA, the new EPAS (CSWE, 2015) makes this obligatory. Coates' experience with the introduction of structural social work and spirituality into the social work curriculum shows that such transformation takes place over several years. A change of this dimension needs an advocate, usually a professor passionate about the topic, who introduces it as something that needs to be considered, and proceeds to prepare a course, write about and even develop a network to discuss and debate the issue at conferences and in academic publications. Several writers have written about the importance of interdisciplinary collaboration in this endeavour (e.g. Norton, 2012; Schmitz, Matyók, Sloan, & James, 2012). Over time, usually with much discussion and debate, and increasing attention in scholarly literature, its relevance becomes clearer and like-minded scholars start to integrate material into their teaching and research. A critical point is reached when foundational or theory courses start to include content to support the new concept such that, eventually, it becomes a foundation itself and begins to influence how the curriculum is structured and taught. In social work, areas as diverse as evidence-based practice and spirituality have become embedded in this way, with varying emphases across national contexts. Jones (2013) refers to the way in which new areas may be added as the bolt-on, embedding and transformative options, and offers an example of transformative education in ecological literacy by bringing community spaces into the social work classroom (Jones, 2014).

However, certain areas, like decolonized or Indigenous social work, and now environmental social work, have been a tough sell as they differ substantially from the foundational assumptions of much existing social work scholarship, especially where it is heavily dominated by Anglo-American scholars rooted in individualistic therapeutic approaches. The value of therapeutic approaches notwithstanding, there are many contexts where a broader relational perspective that situates individual problems in the larger social context of *inter alia* poverty, inequality, patriarchy, classism, racism, heterosexism, privilege and ableism is preferable (Folgheraiter, 2004). Progressive scholars and activists within social work have pushed the profession to examine the personal–political dimensions and structural realities of people's lives and, in so doing, have expanded the boundaries of social work theory and practice (Gray & Webb, 2013a, 2013b). Their success in so doing has possibly led to efforts to better integrate micro and macro practice and to begin to take on a more collective perspective in the new EPAS (Council for Social Work Education [CSWE], 2015) revision in US social work.

Environmental literature has progressed to include a critique of the foundational beliefs informing Western society, and Western social work (Gray, Coates, & Hetherington, 2013). As Mezirow (1991, 1997) showed, perspective transformation begins when core values are challenged, when long-held beliefs and assumptions no longer seem to fit. We have seen such transformations on a broad scale in the way in which people think about the position of women in society, such that it is unconscionable to entertain the idea that women do not have a right to vote. There are numerous examples, such as the way we think about people who identify as lesbian, gay, bisexual, transgender, intersex or queer, even though these groups are still subjected to discrimination in many quarters (Thompson, 2012). There is a tide of change at work that supports the fundamental change of values a holistic environmental perspective entails. For example, the spirituality, Indigenous and decolonization discourses question the dominance of Western values. Today, people think differently about colonialism and the role of missionaries in spreading Christianity and destroying local traditions, cultures, values, languages and beliefs. Decolonization is affecting social work in diverse ways and is best seen in relation to cultural relevance and appropriateness (Gray, Coates, Yellow Bird, & Hetherington, 2013). This is especially true in non-Western contexts and may not resonate with the experience of social workers in North America and Europe, yet.

This changing tide has created greater awareness of climate change and environmental destruction, of gross inequality and the concentration of wealth in the hands of the few, and of the destructive force of corporate greed and excess (Ife, 2013; Klein, 2014), and this is contributing to gradual perspective transformation. It is leading to a critical questioning of the core values of Western society and is seen to be influencing social work through emerging areas of scholarship. Just as feminist thought highlighted patriarchy and male dominance and radical social work the impact of race and class, so holistic environmental thought focuses on the excesses of modernity and its extractive economy, its impact on climate change, and the limitations of a humanistic focus.

Shifting Gears in Our Thinking: Transforming the Curriculum

So what then does this change in perspective mean for social work? First, it suggests the need for a fundamental rethinking of the humanistic values and theories informing social work. Ecocentric environmental values relating to *inter alia* conservation, degrowth, diversity, sustainability, spirituality and restoration can inspire a reformulation of social work's existing anthropocentric values (Gray & Coates, 2012). This requires rethinking social justice in social work; for example, Besthorn (2013) advocates a radical equalitarian ecological approach to justice, which moves beyond the priority of human interests to appreciate the interdependence of all people and the larger body of life, including non-sentient beings and natural systems.

Secondly, it suggests a broad encompassing theoretical framework that incorporates an understanding of environmental issues and their impact, including species extinctions, pollution, habitat destruction, industrial 'accidents' and natural disasters (Dominelli, 2013; Gray et al., 2013). Locating this in the history of the environmental movement, with examples of the role other professions have played, enhances the relevance of these issues for social work. Thus, it opens a terrain for interdisciplinary and interprofessional work most-often called for in environmental practice relating *inter alia* to post-disaster interventions, food security, peace studies, crime prevention and environmental restoration (see Coates & Besthorn, 2010; Dominelli, 2013; Norton, 2012; Schmitz et al., 2013). Also, relating these issues to broader discourses of neoconservatism and globalization shows students how political, economic and social processes produce environmental impacts. It is not just the environment *per se* that we are concerned with, but how our ideas are shaped, so we see structural issues as distinct from, and of lesser concern than, depression or violence, for example. Social work can become part of the growing movement to create a social and political context where social transformation is possible. Hence it suggests a framework for analysing multiple oppressions arising from inter-related social, economic and environmental problems (Gray et al., 2013). Most importantly, it draws attention to the fact that poor people are most affected by environmental destruction and climate change (Hetherington & Boddy, 2013). Here, course material could focus on climate injustice highlighting how climate change impacts on small nations with low (or almost zero) carbon footprints, or food security enabling community gardens, community-supported agriculture or local food purchasing coops to provide people on low incomes with fresh and healthy food (rather than primarily eating food that travels thousands of kilometres). There is a space here for social workers to work as facilitators of active citizenship and advocacy on a local and global scale.

Thirdly, environmental practice highlights a macro role for social workers in, for example, conducting community assessments where environmental impacts are a consideration; using local and 'traditional knowledge' of the environment as an asset on which to design programs, especially in disaster and climate change work; and encouraging foundations and community funding agencies to incorporate environmental concerns as a significant factor in funding applications. Jones (2014) provides an example of a classroom workshop in which students reflect on the nature and extent of social and environmental issues in authentic community-based scenarios; employ systems thinking to examine the connections between these; and design a strategy that simultaneously addresses both these issues grounded in the principles and practices of an ecologically oriented community development approach, which values process, participation and local wisdom.

Finally, the environment offers excellent material for policy courses, especially as an example of the role of political and economic institutions in supporting policies and practices that negatively impact upon the environment or get in the way of environmental regulations and safeguards. It also offers an excellent opportunity to bring in an international perspective showing how international development aid organizations have colluded to promote an extractive economy and the destruction of major ecosystems, and in the process, to exacerbate gross inequality.

When one marries the underlying theoretical approaches and issues, discussed above, with the diverse contexts in which social work is practised, an almost full curriculum begins to emerge if taught along with generalist skills and practice approaches at the individual, group, organizational, community and policy levels. The new Educational Policy and Accreditation Standards of the U.S. Council on Social Work Education (CSWE, 2015) has introduced a new competency that supports the introduction of curriculum content on environmental justice. Social work's leading international organizations have also provided a framework for environmental education and practice in the Global Agenda for Social Work and Social Development (IASSW/IFSW, 2012).

The focus on specific environmental issues needs to grab students' attention. Relating climate change to its impact on people and animals, rather than spending too much time on complex science, especially with video footage, is likely to be most engaging. Students will more easily grasp a holistic environmental perspective if it is presented as an extension of ecological and systems theory that shares the concepts of interdependence, unity, holism and connectedness. An environmental perspective also resonates with evidence-based practice and highlights the contradictions in scientific evidence (Gambrill, 2014). A discussion of research for and against climate change, for example, raises awareness of the uses and misuses of research to support various perspectives. If spirituality or Indigenous social work were already taught in the curriculum, students would more easily grasp how the values underlying an environmental perspective lead to a questioning of long-held beliefs and assumptions (Coates, Gray, & Hetherington, 2006; Gray, Coates, & Hetherington, 2007).

Conclusion

In this paper, we have presented some ideas for the development of a curriculum and the theoretical ideas undergirding such a change. In so doing, we have attempted to add to existing literature and the work of writers interested in advancing an environmental perspective in social work theory, practice and education. The economic excesses and levels of inequality in contemporary society are unsustainable and, as this realization grows in time, interest in the issues that we have raised will grow and perspective transformation will take place, making it impossible for social work to avoid incorporating environmental content, such as that we have suggested, in the curriculum. The next step in promoting environmental justice in social work has already begun, with the literature growing in this area and environmental themes now incorporated in international and local social work conferences. As more and more academics and students, and practitioners, engage in discussions surrounding environmental issues and the role of social work, a climate will be created for new courses in the curriculum and so social work will begin to embrace a holistic environmental perspective and the new practices that emerge therefrom.

Disclosure statement

No potential conflict of interest was reported by the authors.

References

Berry, T. (1988). *The dream of the earth.* San Francisco, CA: Sierra Club Books.

Berry, T. (1999). *The great work: Our way into the future.* New York, NY: Bell Tower.

Besthorn, F. H. (1997). *Reconceptualizing social work's person-in-environment perspective: Explorations in radical environmental thought* (Unpublished doctoral dissertation, University of Kansas, Lawrence).

Besthorn, F. H. (2003). Radical ecologisms: Insights for educating social workers in ecological activism and social justice. *Critical Social Work: An Interdisciplinary Journal Dedicated to Social Justice*, *3*, 66–106. Retrieved from http://www.criticalsocialwork.com/CSW_2003_1.html

Besthorn, F. H. (2012). Deep ecology's contributions to social work: A ten-year retrospective. *International Journal of Social Welfare*, *21*, 248–259.

Besthorn, F. H. (2013). Radical equalitarian ecological justice: A social work call to action. In M. Gray, J. Coates, & T. Hetherington (Eds.), *Environmental social work* (pp. 31–45). Abingdon: Routledge.

Besthorn, F. H., & Canda, E. (2002). Revisioning environment: Deep ecology for education and teaching in social work. *Journal of Teaching in Social Work, 22*, 79–101.

Bronfenbrenner, U. (1979). *The ecology of human development: Experiments by nature and design*. Cambridge, MA: Harvard University Press.

Brown, P. G., & Garver, G. (2009). *Right relationship: Building a whole earth economy*. San Francisco, CA: Berrett-Koehler Publishers.

Capra, F. (1982). *The turning point*. New York, NY: Simon and Shuster.

Clark, C. L., & Asquith, S. (1985). *Social work and social philosophy: A guide to practice*. London: Routledge and Kegan Paul.

Coates, J. (1994). Education for social transformation. *Journal of Teaching in Social Work, 10*(1/2), 1–18.

Coates, J. (2003). *Ecology and social work: Toward a new paradigm*. Halifax, NS: Fernwood Press.

Coates, J. (2005). Environmental crisis: Implications for social work. *Journal of Progressive Human Services, 16*, 25–49.

Coates, J., & Besthorn, F. H. (2010). Introduction—Building bridges and crossing boundaries: Dialogues in professional helping. *Critical Social Work, 11*(3), 1–7.

Coates, J., & Gray, M. (2012). The environment and social work: An overview and introduction. *International Journal of Social Welfare, 21*, 230–238. doi:10.1111/j.1468-2397.2011.00851.x

Coates, J., Gray, M., & Hetherington, T. (2006). An 'ecospiritual' perspective: Finally, a place for indigenous approaches. *British Journal of Social Work, 36*, 381–399.

Council for Social Work Education. (2015). *EPAS revision*. Retrieved from http://www.cswe.org/Accreditation/EPASRevision.aspx

Dominelli, L. (2012). *Green social work*. Bristol: Policy Press.

Dominelli, L. (2013). Global education for disaster relief work. In M. Gray, J. Coates, & T. Hetherington (Eds.), *Environmental social work* (pp. 280–297). Abingdon: Routledge.

Dylan, A. (2013). Environmental sustainability, sustainable development, and social work. In M. Gray, J. Coates, & T. Hetherington (Eds.), *Environmental social work* (pp. 62–87). Abingdon: Routledge.

Folgheraiter, F. (2004). *Relational social work: Toward networking societal practices*. London: Jessica Kingsley.

Gambrill, E. (2014). *Critical thinking in clinical practice* (3rd ed.). Hoboken, NJ: John Wiley & Sons.

Germain, C. B. (1981). The ecological approach to people-environmental transactions. *Social Casework, 62*, 323–331.

Germain, C. B. (1991). *Human behaviour in the social environment: An ecological view*. New York, NY: Cambridge University Press.

Germain, C., & Gitterman, A. (1980). *The life model of social work practice*. New York, NY: Columbia University Press.

Gray, M., & Coates, J. (2012). Environmental ethics for social work: Social work's responsibility to the non-human world. *International Journal of Social Welfare, 21*, 239–247. doi:10.1111/j.1468-2397.2011.00852.x

Gray, M., Coates, J., & Hetherington, T. (2007). Hearing Indigenous voices in mainstream social work. *Families in Society, 88*, 53–64.

Gray, M., Coates, J., & Hetherington, T. (Eds.). (2013). *Environmental social work*. Abingdon: Routledge.

Gray, M., Coates, J., Yellow Bird, M., & Hetherington, T. (Eds.). (2013). *Decolonizing social work*. Farnham: Ashgate.

Gray, M., & Webb, S. A. (Eds.). (2013a). *Social work theories and methods* (2nd ed.). London: Sage.

Gray, M., & Webb, S. A. (Eds.). (2013b). *The new politics of social work*. Basingstoke: Palgrave.

Hannis, M. (2015). The virtues of acknowledged ecological dependence: Sustainability, autonomy and human flourishing. *Environmental Values, 24*, 145–164. doi:10.3197/096327114X13947900181437

Hart, M. A. (2004). *Seeking mino-pimatisiwin: An aboriginal approach to helping* (2nd ed.). Halifax, NS: Fernwood Press.

Hayward, R. A., Miller, S. E., & Shaw, T. V. (2013). Social work education on the environment in contemporary curricula in the USA. In M. Gray, J. Coates, & T. Hetherington (Eds.), *Environmental social work* (pp. 246–259). Abingdon: Routledge.

Hetherington, T., & Boddy, J. (2013). Ecosocial work with marginalized populations: Time for action on climate change. In M. Gray, J. Coates, & T. Hetherington (Eds.), *Environmental social work* (pp. 46–61). Abingdon: Routledge.

Hoff, M. D. (1992). *Integrating global environmental issues into the social work curriculum*. Paper presentation at the annual meeting of the International Association of Schools of Social Work, 1992 conference, Washington, DC.

IASSW/IFSW. (2012). *Global agenda for social work and social development*. Retrieved from http://www.globalsocialagenda.org/

Ife, J. (2013). *Community development in an uncertain world*. Cambridge: Cambridge University Press.

Jones, P. (2008). Expanding the ecological consciousness of social work students: Education for sustainable practice. In *Sustainability in higher education: Directions for change. Proceedings of the EDU-COM 2008 conference* (pp. 285–292). Khon Kaen: EDU-COM.

Jones, P. (2010). Responding to the ecological crisis: Transformative pathways for social work education. *Journal of Social Work Education, 46*, 67–84.

Jones, P. (2013). Transforming the curriculum: Social work education and ecological consciousness. In M. Gray, J. Coates, & T. Hetherington (Eds.), *Environmental social work* (pp. 213–230). Abingdon: Routledge.

Jones, P. (2014). Ecological literacy in social work education: Using a scenario-based approach to bring community spaces into the classroom. *Proceedings of the 11th international transformative learning conference*, New York. Retrieved from http://researchonline.jcu.edu.au/35562/6/35562%20Jones%202014.pdf

Kahn, M., & Scher, S. (2002). Infusing content on the physical environment into the BSW curriculum. *The Journal of Baccalaureate Social Work, 7*(2), 1–14.

Klein, N. (2014). *This changes everything*. Toronto, ON: Knopf.

Korten, D. (2009). *Agenda for a new economy: From phantom wealth to real wealth*. San Francisco, CA: Berrett-Koehler.

Mary, N. (2008). *Social work in a sustainable world*. Chicago, IL: Lyceum Books.

Meyer, C. (Ed.). (1983). *Clinical social work in the ecosystems perspective*. New York, NY: Columbia University Press.

Mezirow, J. (1991). *Transformative dimensions of adult learning*. San Francisco, CA: Jossey-Bass.

Mezirow, J. (1997). Transformative learning: Theory to practice. *New Directions for Adult and Continuing Education, 74*, 5–12.

Mezirow, J. (2012). Learning to think like an adult: Core concepts of transformative theory. In E. Taylor, P. Cranton, & Associates (Eds.), *The handbook of transformative learning* (pp. 73–95). San Francisco, CA: Jossey-Bass.

Norton, C. (2012). Social work and the environment: An ecosocial approach. *International Journal of Social Welfare, 21*, 299–308. doi:10.1111/j.1468-2397.2011.00853.x

Pandey, S. (1998). Women, environment, and sustainable development. *International Social Work, 41*, 339–355.

Rogge, M. E. (1993). Social work, disenfranchised communities, and the natural environment: Field education opportunities. *Journal of Social Work Education, 29*, 111–120.

Schmitz, C. L., Matyok, T., James, C. D., & Sloan, L. M. (2013). Environmental sustainability: Educating social workers for interdisciplinary practice. In M. Gray, J. Coates, & T. Hetherington (Eds.), *Environmental social work* (pp. 260–279). Abingdon: Routledge.

Schmitz, C. L., Matyók, T., Sloan, L., & James, C. D. (2012). The relationship between social work and environmental sustainability: Implications for interdisciplinary practice. *International Journal of Social Welfare, 21*, 278–286. doi:10.1111/j.1468-2397.2011.00855.x

Spretnak, C. (1997). *The resurgence of the real: Body, nature and place in a hypermodern world.* Harlow: Addison-Wesley.

Thompson, N. (2012). *Anti-discriminatory practice: Equality, diversity and social justice.* London: Palgrave Macmillan.

Trainer, T. (1985). *Abandon affluence*. London: Zed Press.

Trainer, T. (2010). *The transition to a sustainable and just world.* Sydney, NSW: Envirobook.

Zapf, M. K. (2009). *Social work and the environment: Understanding people and place.* Toronto, ON: Canadian Scholars' Press.

Sustainable Social Work: An Environmental Justice Framework for Social Work Education

Samantha Teixeira & Amy Krings

Environmental degradation is not experienced by all populations equally; hazardous and toxic waste sites, resource contamination (e.g., exposure to pesticides), air pollution, and numerous other forms of environmental degradation disproportionately affect low income and minority communities. The communities most affected by environmental injustices are often the same communities where social workers are entrenched in service provision at the individual, family, and community level. In this article, we use a global social work paradigm to describe practical ways in which environmental justice content can be infused in the training and education of social workers across contexts in order to prepare professionals with the skills to respond to ever-increasing global environmental degradation. We discuss ways for social work educators to integrate and frame environmental concerns and their consequences for vulnerable populations using existing social work models and perspectives to improve the social work profession's ability to respond to environmental injustices. There are significant social work implications; social workers need to adapt and respond to contexts that shape our practice, including environmental concerns that impact the vulnerable and oppressed populations that we serve.

The social work profession utilizes the 'person in the environment' (PIE) perspective to understand individual and community level problems. However, the profession has largely defined this environmental perspective solely through the lens of the social environment, despite a great deal of knowledge that the built and natural environments are related to health and well-being (Coates & Gray, 2012; Kemp, 2011; McKinnon, 2008; Miller, Hayward, & Shaw, 2012; Rogge, 1993; Zapf, 2009).

There is a growing push for social workers to better understand the interdependence between people and their socio-cultural, economic, and physical environments (Dominelli, 2012; Hoff & Rogge, 1996; Kelley, 2011). These environmentally focused social workers have highlighted the profession's role in simultaneously promoting environmental and social justice.

In this article, we describe practical ways in which social work educators can infuse environmental justice principles and interventions into traditional social work training; recognizing the intersectionality of social and environmental inequality. We first provide a brief background to describe global environmental degradation, with an emphasis on how the distribution of environmental hazards disproportionately burdens the poor and racial minorities. Next, we present an original framework that integrates global social work standards with environmental justice principles. This framework, based on global standards for social work set forth by The International Federation of Social Workers (IFSW) and the International Association of Schools of Social Work (IASSW), includes four practice guideposts with illustrative case examples that can be used by educators to train social work students and practitioners. We conclude with a description of social work implications, in particular, the unique opportunity for social work professionals to bring our skills and perspectives to bear on environmental justice issues. Finally, we recommend that social workers adapt and respond to contexts that shape our practice, integrating environmental concerns that impact the vulnerable and oppressed populations that we serve.

Background

The issue of environmental degradation has seen increased global recognition among social service and health professionals after being highlighted as one of the United Nations' (UN) Millenium Development Goals (United Nations, 2000). As part of this goal, the UN aims to support sustainable development policies and programs that stem the degradation of environmental resources including safe drinking water, firewood, and basic sanitation.

Research suggests that environmental degradation does not impact all populations equally; low income and minority communities are more likely to be exposed to air, land, and water contamination, and are least equipped to mitigate the resulting harm to human health and the natural environment (Brown, 1995; Burger & Gochfeld, 2011; Mohai, Pellow, & Roberts, 2009). In part, this is why low-income racial minorities are more likely to live and work near toxic and hazardous facilities such as interstates, garbage facilities, and wastewater treatment plants (Austin & Schill, 1991; Mohai & Bryant, 1992). These types of noxious facilities – known as 'locally undesirable land uses' or 'LULUs' – appear necessary to the broader society, but create environmental burdens and ultimately health impacts among host community residents.

These land use decisions culminate in a phenomenon known as 'environmental racism' because communities with higher percentages of racial minorities are more likely to be exposed to pollution and toxic waste, bear the brunt of the mental and physical health hazards resulting from environmental degradation, and experience unique vulnerabilities (e.g., powerlessness, oppression) that leave them exposed to environmental hazards (Chavis, 1994; Holifield, 2001; Mohai & Bryant, 1992; Mohai et al., 2009). Because the health of host community residents is sacrificed while the broader society benefits, these places have described as 'sacrifice zones' (Lerner, 2010). The manner in which we distribute environmental benefits and sacrifice zones has serious implications for people's physical and social well-being. Many global conflicts can be traced to issues of access to land, water, oil, and other natural resources.

Environmental sustainability and environmental justice are two principles that can be applied to begin addressing these disparities. Sustainability is broadly defined as using natural resources necessary for human survival responsibly so that they will remain available for future generations (EPA, 2014). Environmental justice interventions aim to promote a safe, clean environment and meaningfully involve all people in policy and development decisions that affect their environment (EPA, 2014; Mohai & Bryant, 1992; Rogge, 1993). Social workers, traditionally concerned with the promotion of human rights and social justice, are primed to be strong partners in environmental justice movements.

Addressing environmental degradation is not a new area for social work but a return to the roots of the profession—a focus that was marginalized as the scope of social work education and practice was narrowed following the professionalization of Flexner's medical model in the early twentieth century. In her call to action urging social workers to return to their environmental roots, Kemp (2011) argued that, 'Although social work is optimally positioned to respond to the human implications of environmental change and stress, the profession largely lacks a presence in environmental practice, research, and policy-making' (p. 1198). The communities most affected by environmental injustices are the same communities where social workers are entrenched in service provision at the individual, family, and community level (McKinnon, 2008). In this article, we join a vocal group of social work educators (Coates & Gray, 2012; Dominelli, 2012; Kemp, 2011; Rogge, 1993) in urging social workers to consider a paradigm shift that embraces the role social workers can play in environmental and social justice advocacy.

Global Standards for Social Work Education and Practice: A Framework to Apply Environmental Justice Content

The International Federation of Social Workers (IFSW) and the International Association of Schools of Social Work (IASSW) jointly developed global standards for social work including an internationally accepted definition of social work and educational practice standards. The creators of the document set forth to identify universal aspects of social work that could guide the training and education of social workers across the globe (IFSW, 2012). They argued that although local context is important in social work training, the core purposes and universal paradigm that guide social workers cut across national boundaries.

Utilizing the paradigm as laid out by the IASSW and the IFSW, this article offers practical suggestions for social work educators to weave environmental justice content into traditional social work training[1]; integrating environmental concerns and their consequences for vulnerable populations (including individuals, families, and communities across the globe).

The key points of the paradigm will be used as a guide to infuse environmental justice content throughout global social work training and education. The primary paradigmatic guideposts[2] we use are:

1. Recognition of the dignity and worth of all human beings, respect and appreciation for diversity and the assumption, identification, and recognition of strengths and potential of all human beings;
2. Recognition of the interconnectedness among micro, mezzo, and macro systems;
3. The importance of advocacy and changes in socio-structural, political, and economic conditions that disempower, marginalize, and exclude people;
4. Focus on capacity-building and empowerment of individuals, families, groups, organizations and communities through a human-centered developmental approach.

Though the authors are from the USA and their experiences reflect US-based training and practice, this article is relevant to a global social work education audience because it focuses on skills, values, and examples of interventions that can be taught in relation to each of the global paradigmatic guideposts outlined above. We focus on core social work values and their application to environmental justice relevant skills and competencies and provide case studies for each guidepost that can be used by social work educators to illustrate how these skills and values can be applied in an international context.

Guidepost 1: Recognition of Dignity and Worth, Diversity, and Strengths Perspective

Social workers who are attentive to social justice examine their interventions through principles of equality and fairness. They commit to respecting the dignity and worth of the individuals with whom they work, regardless of race, social status, or access to power. This lens – utilized to promote social justice – also provides a strong framework for delivering environmental justice content in social work education (Pillai & Gupta, 2012).

Values and Perspectives

Global environmental issues, such as climate change, cut across political, economic, cultural, and social boundaries. Though toxic risk and exposure to pollution and contamination are global phenomena, they disproportionately affect members of low-income and minority communities (Mohai et al., 2009; Morello-Frosch & Jesdale, 2006).

As social work educators, we train our students to identify the root causes of social problems, rather than to 'blame the victim'. This approach, unique to social work, can be applied to issues including environmental degradation. Instead of blaming the poor for their problems, we can emphasize the underlying social, political, and economic systems that produce environmental degradation and the role of social workers in intervention. For example, limited financial options constrain housing options, requiring some poor families to live within hazardous environments (Austin & Schill, 1991). In addition, because it is understood that new hazardous developments will be opposed by any host community, corporate leaders are incentivized to build within politically marginalized areas.

In their recommendations to the Council on Social Work Education (CSWE) Global Commission, Pillai and Gupta (2012) stated that social workers should focus on both social inequality and poverty as the underlying causes of ecological degradation. If we train social work professionals to understand environmental degradation as a problem that intersects with poverty and other problems we commonly address as social workers, we can train them to use our existing skillsets and models of intervention to simultaneously address environmental degradation. As social workers, when we respect the right of all people to a safe, clean environment, we also respect their right to self-determination. The poor are not typically afforded a voice in determining land use and thus may become the default for toxic site location, illegal dumping, or other environmental degradation (Bullard, Mohai, Saha, & Wright, 2008). Social workers have a role to play in ensuring that these traditionally disenfranchised populations have a voice in decisions that affect their individual and environmental well-being.

Skills

One of the key roles of social workers around the globe is to recognize strengths in diverse and oppressed groups and harness their potential for action and empowerment. As part of our training, we prepare students to do this work by developing skills including the ability to demonstrate self-awareness of personal bias (IFSW, 2012; NASW, 2008). Though it is common to work with students to develop awareness of biases such as racial or socioeconomic prejudice, we can expand this practice to the built and natural environment (e.g., biases against working in blighted neighborhoods in legacy cities or slums in rapidly urbanizing cities in the developing world).

Another skill common to social workers worldwide is the ability to recognize and create awareness of the intersections of forms of oppression in shaping life experiences (IFSW, 2012). In order to infuse environmental justice content into this skill area, instructors can work with students to examine the intersection of poverty and environmental health disparities, with special attention to the role that power and privilege play in both poverty and health disparities. Social work educators may use successful environmental justice movements to showcase the ability of marginalized populations to organize and advocate for safe conditions in their neighborhoods and communities.

Case Study Example

The Toxic Free Neighborhoods project, which took place in the Old Town industrial district of San Diego, CA, USA, exemplifies the values and skills inherent in Guidepost 1. Many of the environmental justice campaigns throughout the USA originated with the goal of reducing asthma and other respiratory illnesses associated with exposure to pollution (Sze, 2007). In the Toxic Free Neighborhoods project, for example, after a team of researchers documented the linkage between the neighborhood's high rate of asthma and concentration of manufacturing facilities, they recruited and trained lay health workers or *promotoras de salud* who were women from the neighborhood with children impacted by asthma. The *promotoras* then interviewed residents about their experiences with pollution, as well as asking them to describe what it is like to live with asthma (Minkler, Garcia, Williams, LoPresti, & Lilly, 2010).

The researchers and *promotoras* then created a public awareness campaign based on their 'statistics and stories' that ultimately catalyzed policy makers to address the pollution. This intervention built upon the strengths and lived experiences of the *promotoras*, particularly their dual roles as mothers and co-researchers. Because the research team respected residents' dignity by amplifying their voices, the mothers whose families had experienced the problem first-hand were afforded a voice in land use decisions related to the health of their children. The campaign resulted in the incorporation of environmental justice principles within San Diego's planning processes as well as a plan to relocate harmful, polluting facilities outside of the neighborhood.

Guidepost 2: Recognition of the interconnectedness among micro, mezzo, and macro systems

Social work's recognition of the multi-systemic nature of social problems makes us unique among helping professions. Although we are leaders in the use of the ecological systems perspective to intervene across multiple systems levels, we remain focused largely on social ecology, minimalizing or even ignoring the importance of natural ecological systems (Coates & Gray, 2012; Zapf, 2009). The second paradigmatic guidepost is an area where we can infuse environmental justice content by utilizing our existing systems perspective and person in environment approach to guide students to an understanding of the complex interplay between humans and the environment.

Values and perspectives

In our use of the person in environment and ecological perspectives, we often artificially separate the social and natural environment, when in all actuality the built, natural, and social environments are intertwined (McKinnon, 2008). As social work educators, we can use our existing ecological framework to guide students to understand the natural and built environment as part of the macro context of practice. Our multi-systemic perspective and ability to work within and across systems is an ideal jumping off point to teach social workers to recognize that environmental injustice is a globalized issue with individual, local, and global implications (Mary, 2008). Further, we can help students recognize the interconnectedness of the natural and social environments and their impact on the well-being of the individuals, groups, and communities with whom we work.

Skills

At the heart of the social work profession is our skill at crossing systemic boundaries to act as advocates and organizers while also playing a professional therapeutic role for individuals and communities (Dominelli, 2012). Social workers can bring this practice skill to bear to address issues related to environmental degradation. At the micro level, social workers can apply the person in environment perspective to their individual therapeutic interventions, with specific attention to the role of the natural and built environment in individual outcomes. At the mezzo level, social workers are well positioned to be the conduit for interdisciplinary approaches to enact systems change; for example, linking urban planning professionals and policy makers to organizations providing human services. At the macro level, social workers are skilled at engaging in consciousness raising and social action interventions to bring to light environmental racism and involve community members in efforts to address environmental inequalities. As social work educators, we can help students learn interventions at multiple systems levels and understand the interconnectedness of these approaches, using evidence to inform their practice.

Case Study example

We illustrate Guidepost 2 in action through this case study of the Movement for the Survival of the Ogoni People (MOSOP), a classic example of a grassroots advocacy effort that addressed environmental and social problems using a multi-systemic intervention. This effort, led by affected residents rather than social service providers, can be used to illustrate how social workers can learn from and partner with local movements that promote social work values. The Ogoni people are a Nigerian minority ethnic group residing in Ogoniland, an area of Nigeria rich with natural resources, including crude oil reserves (Adeola, 2000). The Ogoni people face multiple levels of oppression and experience economic deprivation, economic and political discrimination, and environmental degradation due to their location amidst oil fields and other sought-after natural resources. While the natural resources located in Ogoniland create profit for multinational corporations, the Ogoni people bear the brunt of the hazards produced while extracting these natural resources, without economic benefits (Adeola, 2000). The environment of Ogoniland has been severely devastated by hazardous waste dumping and toxic emissions, further marginalizing the Ogoni people.

Led by Kenule Saro-Wiwa, the MOSOP was formed in 1990 to promote consciousness raising among the Ogoni people, mobilize a collective effort to combat disadvantage, and to use non-violent organizing strategies to address oppression (Adeola, 2000). They held mass demonstrations and lobbied their elected representatives to address the pollution and the violent treatment by the military resulting from their resistance. Over time, the group gained international attention through their advocacy and simultaneously promoted human rights and environmental justice, presenting before the United Nations Commission on Human Rights in 1992. Social work professionals can learn from the Ogoni people and their use of advocacy to hold corporations accountable for the ways that they impact the health and well-being of people who are poor and yet live in resource-rich places.

Guidepost 3: The importance of advocacy and changes in socio-structural, political and economic conditions that disempower, marginalise and exclude people

The profession of social work is committed to advocating for changes in conditions that disempower and marginalize individuals and communities. This integral tenet of social work can be used to conceptualize environmental injustice in relation to systems of oppression that we traditionally train students to recognize and intervene in such as poverty, political disenfranchisement, and segregation.

Values and perspectives

Social workers engage with many clients and communities that suffer from environmental health impacts. Some help to mitigate harm by caring for individuals in need. However, the IFSW (2012) stated that it is also necessary to advocate for changes in the conditions that give rise to environmental injustice – those that disempower, marginalize, and exclude people. We already teach students the concepts of empowerment and show them how to use the strengths perspective in their work, often with a focus on individual client strengths. We can expand the way we integrate these values and perspectives into our teaching by utilizing macro, environmental justice-related examples that draw upon client and community strengths to catalyze community action.

Skills

To accomplish these goals, social work educators train future social workers to recognize and articulate the ways in which political, economic, and social structures oppress some identity groups while enhancing the power and privilege of others. In many programs, this type of analysis is already offered. However, there may be new opportunities to teach these practical skills by offering examples of political decisions and 'non-decisions' (Bachrach & Baratz, 1962) that are made (or ignored) relating to the use of land and water (Auyero & Swistun, 2009; Crenson, 1971; Gaventa, 1980). For example, educators can ground critical analysis within local decisions about transportation, water distribution, housing, waste disposal, energy consumption, and pollution.

Social work educators commonly teach methods including consciousness-raising and social action. In order to use these common traditions to integrate environmental justice content, educators might lead trainings, discussion groups, or even public awareness campaigns to call attention to the disproportionate placement of locally undesirable land uses in poor areas and their impact on health. Students might analyze questions like, 'where are all the trash disposal centers located within the state?' and 'what are the demographics in these neighborhoods?' to better draw the connection between political disenfranchisement, marginalization, and environmental injustice.

Beyond analysis and consciousness-raising, skills relating to mobilization, collective action, and advocacy are necessary. Social workers already act as advocates, community organizers, lobbyists, policy makers, and researchers. We can expand upon these roles to promote environmental justice. For example, Lavelle and Coyle (1992) documented that not only are toxic facilities systematically placed within low-income communities of color in the USA, but they are also fined less often by the Environmental Protection Agency and, when they are fined, receive less serious consequences than similar factories in comparably affluent or white areas. Globally, the issue is even more stark; 95% of individuals affected by pollution-related health problems reside in developing countries and for them, pollution management is an urgent issue (World Bank, 2015). Social workers can assist within existing environmental justice campaigns or mobilize new ones designed to prevent the location of toxic facilities in poor areas while calling for increased accountability. This builds upon our skills relating to recruitment, coalition-building, and lobbying, and can be grounded within an environmental justice perspective.

Case Study Example

Guidepost 3 calls for changes in socio-structural, political, and economic conditions that disempower, marginalise, and exclude people. Our first and second case studies exemplified policy reforms shaped by communities that host locally undesirable land uses (LULUs)—such as factories and oil refineries. In many of the most marginalized communities, environmental justice campaigns have yet to arise. These types of communities provide examples that can be used by social work educators to help students think critically about how to mobilize campaigns that address broader political and economic conditions that may not be easily addressed by fragmented, politically disenfranchised communities. This case study describes political and economic conditions that have led to the disenfranchisement of the Roma people in Europe and how they have led to stark environmental health disparities.

The Roma people are one of the largest minority groups in the European Union (EU); an estimated 10–12 million Roma live across Europe (European Commission, 2015). They have been discriminated against by majority populations for more than 1000 years, first as slaves in Hungary and Romania, then systematically targeted for extermination by Nazis, and today they remain the targets of discriminatory housing and economic policies (Fox, 2001). Due to systematic exclusion from the formal workforce, the Roma are often forced into occupations that involve exposure to highly toxic hazards including coal and uranium mining and scrap metal processing (Antypas et al., 2007; Harper, Steger, & Filčák, 2009).

This political, racial, and economic discrimination also results in geographic marginalization; the Roma are often relocated from towns with strict housing ordinances and placed in highly polluted, undesirable geographic areas. This systemic discrimination means that the Roma are more likely to live in areas proximate to waste treatment facilities, industrial sites, and high traffic areas (Antypas et al., 2007). Research suggests that environmental health disparities contribute to a 10–15 year reduced life expectancy among Roma compared to their non-Roma counterparts (Antypas et al., 2007; Parry et al., 2004). Though Romani activists and Hungarians have been pushing to organize the Roma community to address discriminatory policies that lead to environmental injustices, wide health disparities still exist (Harper et al., 2009). A promising policy framework for environmental health exists in the EU but there is still need for more concrete efforts to promote environmental justice. This is a potential political advocacy opportunity for social workers and environmentalists who may be able to work with Romani activists to address the complex intersection of racial discrimination, exclusion, and poor environmental health.

Guidepost 4: Focus on Capacity-building and Empowerment of Individuals, Families, Groups, Organizations and Communities through a Human-centered Developmental Approach

Social workers view empowerment as both a process and an outcome. We promote interventions and approaches that simultaneously build individual and community capacity, which can allow us to leverage assets across systems. This human-centered and community-centered approach promotes participatory decision making, particularly with regard to environmental issues.

Values and Perspectives

Though the empowerment perspective has roots in numerous academic disciplines, it is among the most commonly used perspectives in social work practice (Gutierrez, Parsons, & Cox, 1998). Despite its fundamental role in social work research and practice, empowerment is often an ambiguous concept. For example, participation in decision-making processes can be an empowering experience when a participant's input is valued and implemented. However, participation can be tokenizing or even manipulative when participants are not taken seriously and do not have influence over the outcome (Arnstein, 1969). At its most basic form, empowerment requires access to information. In the case of environmental injustices, social workers may need to take highly technical and scientific information and translate it into a form that ordinary citizens can understand. Furthermore, they may need to utilize the empowerment perspective to push for more participatory governance, which requires the understanding of various proposals under consideration and the viewpoints all those affected – including government officials, developers, and residents.

Skills

Social workers have historically used empowerment interventions to address numerous macro-level social issues, including environmental injustices. For example, Jane Addams and Hull House staff engaged residents to intervene in neighborhood quality of life issues (e.g., waste disposal, workplace safety) through community organizing, consciousness raising, and other essential social work skills related to empowerment (Kemp, 2011). More recently, global health professionals, including social workers, have utilized a variety of innovative approaches to empower historically oppressed populations to address environmental issues. Social workers can utilize a variety of empowering research and engagement approaches including but not limited to Community Based Participatory Research (CBPR) (Minkler & Wallerstein, 2011; Wallerstein & Duran, 2008), Participatory Action Research (PAR) (Whyte, 1991), community organizing, and consensus building (Fisher & Kling, 1993; Fisher & Shragge, 2000). Social work researchers and practitioners alike can perform this role using long-established social work skills such as engaging, assessing, and intervening within at-risk communities. They can co-produce knowledge about environmental issues and their impacts with local stakeholders, and they can disseminate research data so that it can be used to further the community's visions for change.

Case Study Example

Empowerment is perhaps the most widely used intervention tool in the field of social work (Pillai & Gupta, 2012). Its core philosophy is that people should have influence over the decisions that impact their lives. A well-known example of empowerment in the context of environmental justice is the Dudley Street Neighborhood Initiative in which a Boston community was identified by city officials and developers as a site for urban renewal, to replace affordable, but blighted housing units with market-rate housing (Medoff, 1994). While the residents agreed with the need to address illegal dumping and blight within their community, they were concerned because they were both left out of the planning process and feared that they—like other communities that were bulldozed in the name of urban renewal–would be either forcefully relocated or priced out through gentrification.

In response, the residents of Dudley Street organized using an empowerment approach. They secured funding to hire local organizers to craft environmental justice campaigns, including one titled 'Don't Dump on Us' to shut down illegal dumping of garbage and hazardous materials. In addition, they development their own resident-driven planning process to create new housing that was safe and affordable. They maintained their new development and surrounding environment by organizing a youth council, painting murals, and promoting organized activities in public park spaces (Medoff, 1994). Their use of empowering, participatory strategies can serve as an example for training social workers to critique top-down planning that culminates in environmental disadvantage and, more importantly, to offer alternatives.

Implications for Social Work Education

The framework described in this article has important implications for educating future generations of social workers. We argue that social work not only has the skills and tools to address environmental problems, but, as a discipline our theories and perspectives make us uniquely qualified to work at the intersection of social and environmental justice. Unfortunately, to our knowledge, few social work programs explicitly incorporate environmental justice content into their courses, so social workers may not perceive environmental justice as an issue within the profession's purview (Kemp, 2011).

We need to prepare our students to address this global issue. Social workers have many tools that can help mitigate environmental risk among vulnerable populations. We recognize the importance of self-determination and have the skills to bring marginalized communities to the table in order to change decision-making processes that in the past, have disproportionately harmed already oppressed groups. As the social work profession grapples with the implications of globalization, we will need to better integrate environmental content into our classrooms, our field training, and our daily practice. The framework presented in this article can be used as a first step that will expose students to environmental justice through our existing social work paradigms. Through this exposure, social workers may more readily take their place as interdisciplinary players in the global response to environmental degradation.

Conclusion

Social workers are charged with adapting and responding to contexts that shape our practice (CSWE, 2008). Environmental degradation is a pressing, global problem and is concentrated in oppressed populations and oppressed geographic regions. This matters to our clients. The human impacts of environmental challenges fall most heavily on those to whom social workers are most accountable (Hoff & Rogge, 1996; Kemp, 2011).

In this article, we described ways in which social work educators can infuse environmental justice content in social work education and training while using existing social work perspectives, skills, and interventions. We urge social work educators to tackle the issue of environmental justice by first, educating future social workers to view environmental justice as a core component of social justice, and second, to bring social work values and skills to partnerships addressing environmental degradation. We are not the first to urge the profession to expand the ecological model beyond the social environment to a more holistic view of the social, natural, and built environment as spheres for social work practice; in fact, it is deep within the historical roots of our profession (Dominelli, 2012; Hoff & Rogge, 1996; Kemp, 2011; Pillai & Gupta, 2012; Rogge, 1993). Social work education is poised to use our existing practice models to train students to understand that environmental justice *is* social justice.

Disclosure statement

No potential conflict of interest was reported by the authors.

Notes

[1] Because the framework focuses on the core purposes and universal paradigm that guides social work education, this framework is applicable for undergraduate and graduate education.

[2] The original document had 8 epistomological paradigms that we consolidate into 4 environmental justice relevant guideposts.

References

Adeola, F. O. (2000). Cross-national environmental injustice and human rights issues: A review of evidence in the developing world. *American Behavioral Scientist, 43*, 686–706. doi:10.1177/00027640021955496

Antypas, A., Borthwick, F., Cahn, C., Filcak, R., Harper, K., & Steger, T. (2007). Making the case for environmental justice in Central and Eastern Europe. Retrieved from http://www.env-health.org/IMG/pdf/ceu_teljes_pdf.pdf

Arnstein, S. R. (1969). A ladder of citizen participation. *Journal of the American Institute of Planners, 35*, 216–224. doi:10.1080/01944366908977225

Austin, R., & Schill, M. (1991). Black, brown, poor & poisoned: Minority grassroots environmentalism and the quest for eco-justice. *Kansas Journal of Law and Public Policy, 1*, 69.

Auyero, J., & Swistun, D. A. (2009). *Flammable: Environmental suffering in an Argentine shantytown.* New York, NY: Oxford University Press.

Bachrach, P., & Baratz, M. S. (1962). Two faces of power. *American Political Science Review, 56*, 947–952. doi:10.2307/1952796

Brown, P. (1995). Race, class, and environmental health: A review and systematization of the literature. *Environmental Research, 69*, 15–30. doi:http://dx.doi.org/10.1006/enrs.1995.1021

Bullard, R. D., Mohai, P., Saha, R., & Wright, B. (2008). Toxic wastes and race at twenty: Why race still matters after all of these years. *Environmental Law (Northwestern School of Law), 38*, 371.

Burger, J., & Gochfeld, M. (2011). Conceptual environmental justice model for evaluating chemical pathways of exposure in low-income, minority, native American, and other unique exposure populations. *American Journal of Public Health, 101*, S64–S73. doi:10.2105/AJPH.2010.300077

Chavis, B. F. (1994). Preface. In R. D. Bullard (Ed.), *Unequal protection: Environmental justice and communities of color (pp. xi–xii)*. San Francisco, CA: Sierra Club Books.

Coates, J., & Gray, M. (2012). The environment and social work: An overview and introduction. *International Journal of Social Welfare, 21*, 230–238. doi:10.1111/j.1468-2397.2011.00851.x

Crenson, M. A. (1971). *The un-politics of air pollution: A study of non-decision-making in the cities.* Baltimore, MD: Johns Hopkins Press.

CSWE. (2008). Educational Policy and Accreditation Standards (EPAS): Council on Social Work Education. Retrieved from http://www.cswe.org/File.aspx?id=13780

Dominelli, L. (2012). *Green social work: From environmental crises to environmental justice.* Malden, MA: Polity.

EPA. (2014). What is sustainability? Retrieved from http://www.epa.gov/sustainability/basicinfo.htm

European Commission. (2015). EU and Roma. Retrieved from http://ec.europa.eu/justice/discrimination/roma/index_en.htm

Fisher, R., & Kling, J. (Eds.). (1993). *Mobilizing the community: Local politics in a global era.* Newbury Park, CA: Sage.

Fisher, R., & Shragge, E. (2000). Challenging community organizing: Facing the 21st Century. *Journal of Community Practice, 8*(3), 1–19. doi:10.1300/J125v08n03_01

Fox, J. (2001). Patterns of discrimination, grievances and political activity among Europe's Roma: A cross-sectional analysis. *Journal of Ethnopolitics and Minority Issues in Europe, 2*, 1–25.

Gaventa, J. (1980). *Power and powerlessness: Quiescence and rebellion in an Appalachian Valley (vol. 36)*. Urbana, IL: University of Illinois Press.

Gutierrez, L. M., Parsons, R. J., & Cox, E. O. (1998). *Empowerment in social work practice. A sourcebook*. Pacific Grove, CA: Brooks-Cole.

Harper, K., Steger, T., & Filčák, R. (2009). Environmental justice and Roma communities in Central and Eastern Europe. *Environmental Policy and Governance, 19*, 251–268. doi:10.1002/eet.511

Hoff, M. D., & Rogge, M. E. (1996). Everything that rises must converge: Developing a social work response to environmental injustice. *Journal of Progressive Human Services, 7*, 41–57.

Holifield, R. (2001). Defining environmental justice and environmental racism. *Urban Geography, 22*, 78–90. doi:10.2747/0272-3638.22.1.78

IFSW. (2012). Global standards for the education and training of the social work profession. Retrieved from http://ifsw.org/policies/global-standards/.

Kelley, M. (2011). On the social construction of place: Using participatory methods and digital tools to reconceive distressed urban neighborhoods. In S. E. Sutton & S. P. Kemp (Eds.), *The paradox of urban space: Inequality and transformation in marginalized communities*. New York, NY: Palgrave-Macmillan.

Kemp, S. P. (2011). Recentring environment in social work practice: Necessity, opportunity, challenge. *British Journal of Social Work, 41*, 1198–1210. doi:10.1093/bjsw/bcr119

Lavelle, M., & Coyle, M. (1992). Unequal protection: The racial divide in environmental law. *National Law Journal, 15*, S1–S12.

Lerner, S. (2010). *Sacrifice zones: The front lines of toxic chemical exposure in the United States*. Cambridge, MA: MIT Press.

Mary, N. L. (2008). *Social work in a sustainable world*. Chicago, IL: Lyceum Books, Inc.

McKinnon, J. (2008). Exploring the nexus between social work and the environment. *Australian Social Work, 61*, 256–268. doi:10.1080/03124070802178275

Medoff, P. (1994). *Streets of hope: The fall and rise of an urban neighborhood*. Boston, MA: South End Press.

Miller, S. E., Hayward, R. A., & Shaw, T. V. (2012). Environmental shifts for social work: A principles approach. *International Journal of Social Welfare, 21*, 270–277. doi:10.1111/j.1468-2397.2011.00848.x

Minkler, M., Garcia, A. P., Williams, J., LoPresti, T., & Lilly, J. (2010). Sí se puede: Using participatory research to promote environmental justice in a Latino community in San Diego, California. *Journal of Urban Health, 87*, 796–812. doi:10.1007/s11524-010-9490-0

Minkler, M., & Wallerstein, N. (2011). *Community-based participatory research for health: From process to outcomes*. San Francisco, CA: Jossey-Bass.

Mohai, P., & Bryant, B. (1992). Race, poverty, and the environment. *EPA Journal, 18*, 6–8.

Mohai, P., Pellow, D., & Roberts, J. T. (2009). Environmental justice. *Annual Review of Environment and Resources, 34*, 405–430. doi:10.1146/annurev-environ-082508-094348

Morello-Frosch, R., & Jesdale, B. M. (2006). Separate and unequal: Residential segregation and estimated cancer risks associated with ambient air toxics in U.S. metropolitan areas. *Environmental Health Perspectives, 114*, 386–393. doi:10.1289/ehp.8500

NASW. (2008). *Code of ethics*. Washington, DC: National Association of Social Workers.

Parry, G., Van Cleemput, P., Peters, J., Moore, J., Walters, S., Thomas, K., & Cooper, C. (2004). *The health status of gypsies and travellers in England*. Sheffield: The University of Sheffield School of Health and Related Research.

Pillai, V. K., & Gupta, R. (2012). The greening of social work. Retrieved from http://www.cswe.org/CentersInitiatives/KAKI/50754/66942.aspx

Rogge, M. E. (1993). Social work, disenfranchised communities, and the natural environment: Field education opportunities. *Journal of Social Work Education, 29*, 111–120.

Sze, J. (2007). *Noxious New York: The racial politics of urban health and environmental justice.* Cambridge, MA: MIT Press.

United Nations. (2000). We can end poverty: Millenium development goals and beyond 2015. Retrieved from http://www.un.org/millenniumgoals/environ.shtml

Wallerstein, N. B., & Duran, B. (2008). The theoretical, historical, and practice roots of community based participatory research. In M. Minkler & N. Wallerstein (Eds.), *Community based participatory research for health* (pp. 27–52). San Francisco, CA: Jossey-Bass.

Whyte, W. F. (1991). *Participatory action research.* Thousand Oaks, CA: Sage.

World Bank. (2015). Pollution management and environment health. Retrieved from http://www.worldbank.org/en/topic/environment/brief/pmeh

Zapf, M. K. (2009). *Social work and the environment: Understanding people and place.* Toronto, ON: Canadian Scholars' Press.

Social Work and Sustainable Social Development: Teaching and Learning Strategies for 'Green Social Work' Curriculum

Julie Drolet, Haorui Wu, Matthew Taylor & Allyson Dennehy

It is increasingly recognized that social work has a vital role to play in shaping effective responses to environmental degradation, climate change and disasters. Innovative teaching and learning strategies can be used to facilitate understanding of complex, interrelated and systemic social, economic and environmental challenges. This article shares the development and design of a new social work course entitled 'Social work and sustainable social development'. Teaching methods and strategies are provided including individual and group mind maps, case studies and presentations, guest speakers and evaluation methods. Implications for social work education and the profession are discussed in order to build capacity to address environmental justice and sustainability.

Introduction

The profession of social work is being called upon to consider the integration of social, economic and environmental priorities into policies and programs at all levels. The impacts of climate change, disasters, poverty, unemployment, disparities in wealth, violence and gender inequities, among others, contribute to a new imperative to prepare social work students to address these interrelated concerns and their effects on diverse populations at the local and global level (Drolet, 2015). Sustainable development is about meeting the needs of today without compromising the needs of future generations. The concept of sustainability and the implementation of sustainable strategies at local, provincial, national, and international levels are critically important, especially with the severe effects of climate change and disasters on ecosystems and human well-being. Social workers are challenged in their practice to consider complex, interrelated and systemic problems that require new approaches and innovations. This article will share how complex, social, economic and environmental issues were incorporated in a new undergraduate social work course entitled 'Social work and sustainable social development' taught in the semester of Fall 2014 in the Faculty of Social Work at the University of Calgary in Edmonton, Alberta, Canada. The Taylor Institute for Teaching and Learning awarded the instructor a project grant to develop new approaches to learning on the role of social work and sustainable social development. A teaching initiative project aimed to advance teaching and learning practices centred on the integration of social, economic and environmental dimensions. The objectives of the project are (1) to envision new teaching approaches and innovations appropriate for complex, interrelated and systemic problems; (2) to design teaching methods in consultation with practitioners that facilitate student understanding of the role of social work in sustainable social development; (3) to develop visual teaching tools to foster student learning on complex issues; (4) to integrate participatory learning activities in the classroom using case studies developed by students for students and (5) to conduct formative and summative evaluations of new teaching approaches and innovations. This article aims to share teaching methods such as mind maps, case studies, guest speakers and evaluation methods with social work educators, academics and practitioners, in order to build capacity within social work education to address environmental justice and sustainability.

Literature Review

The social, economic and environmental challenges facing our world are legion (Mosher, 2010). Frequently global environment crises and their impact on people's physical and social environments have challenged social work practice (Dominelli, 2012), yet linger largely on the periphery of social work discourse, practice and education (Coates, 2005). Ongoing and recurrent crises have exposed the depth of human insecurity and the inadequacy of systemic and structural reforms; this reality contributes to a new climate and urgency for thinking about the social dimensions of development (UNRISD, 2014). McKinnon (2008) deems that the focus on growing environmental concerns such as environmental degradation and climate change has not received significant attention in the social work literature. Jones (2013) maintains that the major social work journals reveal a dearth of articles specifically emphasizing the profession's role in responding to and ameliorating the changing natural environment. Despite the increasing evidence of the many ways in which climate change, environmental degradation and ecological crises are negatively impacting human well-being, social work remains in a nascent position with regards to addressing these challenges and exploring ways to move forward (Jones, 2010; Peeters, 2012). This is particularly true in social work education.

The social work profession has been slow to incorporate the natural or physical environmental as an outside influence that contributes to people's well-being (Zapf, 2012). A study conducted by Jones (2013) concluded that there is little evidence of the inclusion of the natural environmental sustainability in Australian social work curriculum. Another study conducted by Shaw (2013) found that over two-thirds of the respondents (67.56%) stated that their social work education did not include discussions of the natural environment. Shaw (2013) results indicated that the overwhelming majority (90.08%) believed that issues related to social work and the natural environment should be discussed in schools of social work. Social workers are seeking to address the adversity experienced by their clients due to environmental degradation. Social work education should directly address the interrelationship between human beings and the natural environment (Boetto & Bell, 2015). Social work educators and practitioners have an important role to play in developing and sustaining environmental justice in a way that respects the inherent dignity and worth of a person (Dominelli, 2014). The development of this new social work elective course sought to address a critical gap in the social work curriculum.

In a time of intensifying environmental distress, if social work is to maintain its contemporary relevance and value, it must become a profession that incorporates and contributes to an understanding of the changing natural environment and how this is impacting humanity (McKenzie, 2004; McKinnon, 2008). This is especially relevant to social work as climate change and environmental degradation are having, and will continue to have, a disproportionate impact on poor, marginalized and vulnerable populations (McKinnon, 2008). Current social and environmental challenges call for a significant social transition to a sustainable society and sustainable ways of living (Peeters, 2012).

Social workers can make an important and unique contribution to the environmental movement because of their strong skill set in promoting equality, solidarity, human rights and the dignity and worth of all human beings which can be implemented on all levels of practice to promote sustainability (Norton, 2012). Social workers must be at the forefront in the field advocating in partnership with disenfranchised communities (Bent-Goodley, 2015). The 'Global agenda for social work and social development' recognizes a collective commitment to promote social and economic equalities, human rights and dignity, social relationships and environmental sustainability (Global Agenda, 2012). Social workers can use this document when working with vulnerable and disenfranchised populations facing economic, social and environmental issues. Lysack (2015) demonstrates that social work can develop effective policy influencing and environmental advocacy in collaboration with non-governmental organizations (NGOs) as a key social work practice. It is incumbent upon social work education to prepare students for this challenge, and for social work practitioners to embrace sustainability (Hawkins, 2010).

The development and design of the new course begins by acknowledging that social work has a vital role to play in shaping effective responses to social, economic and environmental crises. The focus on sustainable social development aims to achieve this integration (Drolet & Sampson, 2014). The social work profession is uniquely situated to face these challenges due to its historic focus on social systems as well as its advocacy-based and action-oriented framework for practice (Drolet, Samuel, Larson, Tesoriero & Ginsberg, 2010; Hawkins, 2010). Environmental issues have had little attention in the social work profession; nonetheless, this is changing rapidly as environmental concerns, specifically climate change and disasters, impact people's health, livelihoods, social networks, vulnerability and risk levels (Alston & Whittenbury, 2012; Dominelli, 2011, 2012; Drolet, 2014; Heinonen & Drolet, 2012; Drolet, Samuel & Larson, 2012; Ersing & Kost, 2012).

Green Social Work and Social Work Education

Dominelli (2012) *Green Social Work* focuses on how responses to environmental crises must both challenge and address poverty, structural inequalities, socio-economic disparities, industrialization processes, consumption patterns, global interdependencies and limited natural resources. The concept of green social work provides a holistic understanding about the various environments and their impacts upon people's behaviour. It is widely recognized that teaching on the topic of sustainability is a complex and challenging undertaking that involves exploring multiple, interconnected and overlapping elements of social, economic and environmental challenges (Dale & Newman, 2005; Matthias Barth, Godemann, Rieckmann, & Stoltenberg, 2007). The green social work approach was particularly useful in examining environmental issues from a social work perspective, and the text provides examples on how social workers can engage in community initiatives. The topic of sustainability is of growing interest to social work students, social work faculty, and social work practitioners and is a critically important topic for research on teaching and learning. The design and development of the new course has potential for application by educators in other post-secondary institutions on how to integrate complex research evidence and green social work in social work education. The teaching and learning project was guided by the following questions: How is the curriculum to be shaped around sustainable social development? Given the impact of disasters (man-made and/or natural) upon experience, how does this change the teaching and learning experience? How does social work education—on theoretical and pedagogical levels—respond to new challenges emerging from these intersecting relationships? What teaching and learning practices best prepare students for working with marginalized people? These questions are linked to the project goal to advance teaching and learning practices centred on the integration of social, economic and environmental dimensions in social work education and practice so that current and future social worker students can adequately address these complex issues.

For social work educators, the challenge is to equip students with the knowledge and skills to become effective practitioners while also providing opportunities for students to transform abstract concepts into practical actions (Jones, 2013; Steinemann, 2003). The green social work text by Dominelli (2012) provided a theoretical understanding with linkages and applicability to local practice experience. The goal of sustainability-focused education is to move beyond the theoretical and connect classroom instruction to the real world (Brundiers & Wiek, 2011).

An additional goal is to develop the ability of students to be adaptive and flexible in the implementation of their learning as they respond to changing conditions and challenges (Dale & Newman, 2005). The use of problem-based learning techniques has been presented as an effective way to provide these opportunities and skills (Dale & Newman, 2005; Steinemann, 2003). In addition, transformative learning theory has also been recommended (Jones, 2010). Transformative learning theory is about the nature of change and about the processes through which we produce a shift in the way we see and make meaning of the world (Jones, 2010). Both approaches utilize collaborative problem-solving strategies to teach about issues of sustainability, and educators from a diverse spectrum of practices and backgrounds also promote and utilize these strategies (Brundiers & Wiek, 2011; Brundiers, Wiek, & Redman, 2010; Dyer, Selby, & Chalkley, 2006; Kopnina & Meijers, 2014).

Problems pertaining to social, economic and environmental sustainability are complex and require sophisticated responses and extensive problem-solving procedures (Brundiers et al., 2010). The use of problem-solving activities emphasizes learning through application and challenges students to take ownership of their learning experience (Steinemann, 2003). Problem-based learning presents students with challenges that are personally and socially relevant and which are based on real world and real-time problems. In addition, collaborative problem-solving methods provide students with opportunities to develop their professional co-operation and communication skills (Steinemann, 2003).

Teaching Methods and Strategies

The development of the course began with a review of the literature (academic peer-reviewed publications and grey literature) on teaching approaches and innovations that address complex, interrelated and systemic problems. The findings from the literature informed curriculum development, teaching and learning practices. Through a combination of methods, it was possible to adopt a comprehensive and holistic approach to understanding how diverse teaching and learning practices facilitate student learning. The development of this new course was built upon the instructor's research program in the fields of international social work, climate change, disaster social work and experiential learning in social work field education programs. The course sought to facilitate participatory and interactive learning among students in the classroom using innovative methods. Individual and group mind maps, case studies, guest speakers and formative and summative evaluations formed key components of the course. These teaching methods and techniques allowed for learning on the well-being of individuals, families and communities in relation to environmental degradation; practice interventions to address the effects of environmental degradation and the integration of environmental concerns and consequences for vulnerable populations to advance the social work professional in the Canadian and international context. Research ethics approval was obtained by the University of Calgary's research ethics board prior to engaging in consultations with practitioners on sustainable development and teaching and learning strategies. Students were asked to sign a letter of permission in order to reproduce individual and group mind maps developed during the course in a variety of print and digital formats.

The learning objectives of this course are

(1) To define sustainable social development and understand the importance of integrating social, economic and environmental dimensions in social work practice and policy.
(2) To critically examine the concept of 'development' and globalization at the local, provincial, national and international levels.
(3) To consider 'green social work' and the impacts of climate change and disasters.
(4) To identify issues surrounding vulnerability, resilience, sustainability and environmental justice.
(5) To understand the role of social workers in promoting sustainable developsment and sustainability at the local and global levels.

The required course textbook was Dominelli (2012) *Green Social Work*, and a number of readings were included in the course syllabus drawing from academic peer-review journals and book chapters. The assignments for the course included the following: course participation (20%), mind maps (individual 20% and group 20%), case study (30%) and presentation (10%).

Mind Maps

'Mind maps' is an effective tool for teaching students to think conceptually, especially with regards to complex topics or issues, and to understand the interconnectivity and relationships amongst different variables (Trepagnier, 2002). It is a form of active learning, emphasizing organization and exploration, which can lead to a sophistication and synthesis of knowledge of abstract and detailed concepts (Jacobs-Lawson & Hershey, 2002). The development of 'mind maps', as a teaching tool for visualizing the whole concept and/or issue, challenges students to envision the greater context in which a concept exists (Trepagnier, 2002). Specifically, 'mind maps' were used to present new ideas and information, to illustrate the links and connections between levels and units and to encourage students to grasp complex information (Bochel, n.d.). The use of mind maps is an innovative way of facilitating learning to understand key information, and provides students with a strategy for retaining information and integrating critical thinking and problem-solving skills (Noonan, 2013). 'Mind maps' promote active learning and assist students to use their full cognitive abilities, improve memory and learning skills, allow students to share ideas and benefit the needs of different learning styles (Peterson & Snyder, 1998).

Mind maps can be developed and used by individuals and in groups, and the creation of a mind map challenges students to explore, in detail, their understanding of a certain concept and the relationships amongst associated variables (Jacobs-

Lawson & Hershey, 2002). Students were provided with the following information on developing mind maps:

- Mind maps visually represent hierarchical information that includes a central idea surrounded by connected branches of associated topics.
- Step 1—Create a central idea—the central idea is the starting point of the mind maps and represents the topic to be explored. The central idea should be in the centre of the page and be associated with an image.
- Step 2—Add branches to the map—the main branches which flow from the central image are the key themes. Each theme or branch can be explored in greater depth by adding smaller branches.
- Step 3—Add keywords—an important principle of the mind maps is to use one word per branch. A variety of keywords can be used to explore key themes.
- Step 4—Colour code the branches—the use of colour in the mind maps creates links to categorize, highlight, analyse information and identify connections. Students were asked to use three or more colours in their mind maps. Various art supplies were brought to class by the instructor for this purpose (e.g. coloured markets, highlighters, etc.)
- Step 5—Include images—the use of images can convey more information in mind maps.

Individual Mind Maps

The instructor provided a lesson on mind maps and shared resource material on the development and design of mind maps in social science disciplines. Students were assigned individual mind maps in order to gain an understanding of how their ideas could have been organized. The aim of doing mind maps was to link course-related materials together in a creative way so that it would retain learning and express creativity. Figure 1 is an example of an individual mind map developed by a student, which was presented in class to facilitate peer learning and a more in-depth understanding of the key themes, concepts and relationships.

Individual mind map on sustainability

In Figure 1, the mind map's central theme is 'sustainability'. The four branches portrayed represent the environment, agriculture, economy and community, and these branches are related to the central theme of 'sustainability'. All four branches relate to sustainability in how one can achieve it, how one contributes to it, how one's sustainability is affected by larger systemic forces and how sustainable development requires collaboration. The student demonstrated an understanding of sustainable development by choosing the branch word 'community'. The student also demonstrated an understanding of sustainability by choosing environment, economy and community as the expanding concepts, representing the three pillars for sustainable development and green social work by displaying the need to protect the Earth. The green social work perspective integrates preserving the Earth's resources for current and future generations (Dominelli, 2012). The student used the term 'fair

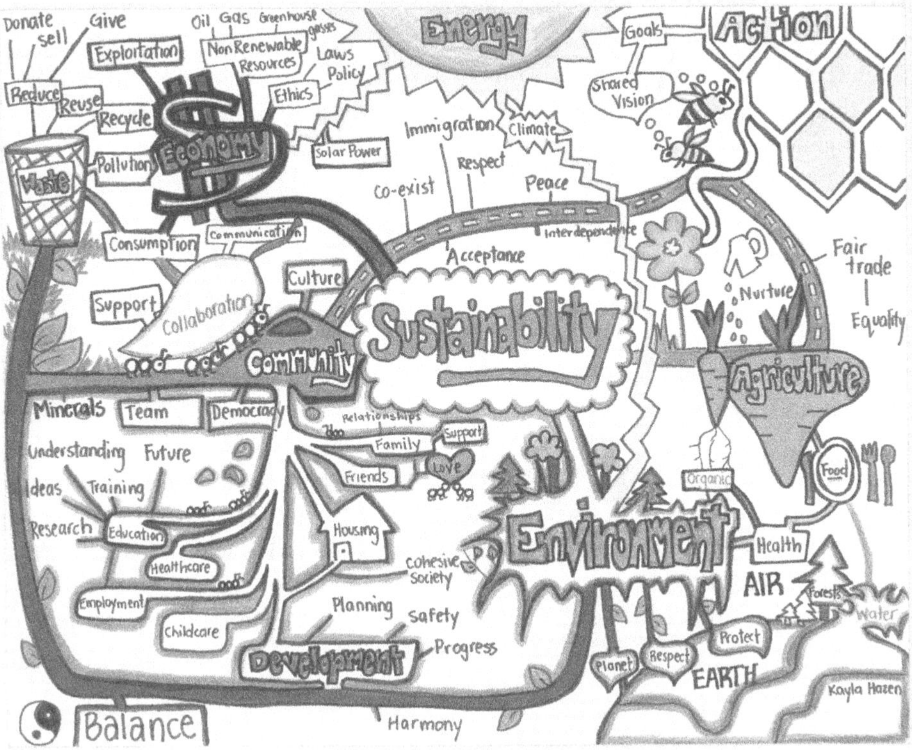

Figure 1 Individual Mind Map on Sustainability Figure: Group Mind Map on Resilience.

trade' because it challenges poverty and structural inequalities. Social development is represented by the terms development, progress, community and support. The mind map reflects the social development concept by linking development as a process that requires community support to improve the quality of peoples' lives. The mind map enriched students' understanding of complex and interrelated concepts by organizing and exploring information and relationships in a new way, and by promoting peer learning in the classroom by sharing one's work.

Group Mind Maps

The group mind map assignment was conducted after the individual mind maps assignment in order to build upon student's learning on how to construct and organize a mind map. The group assignment brought together three to four students, in small groups, to learn from each other in brainstorming and critical thinking sessions and discussion groups. The group processes allowed students to consider how their peers might think differently about key environmental issues. The students were challenged to collaborate in a team environment using creative and arts-based methods. A grading rubric was developed and provided to students with the following criteria: (1) Visual (use of a central image; illustrate the links, connections and/or relationships; use of symbols; use of image or graphics or pictures and use of colour to facilitate understanding) and (2) Content (identify a central idea or theme; create 'branches' of related ideas or themes; ability to communicate the idea; use of a key meaningful word per branch and use of varied lines for association). Each group shared their mind maps in class presentations to facilitate student learning and exchanges. Figure 2 provides a mind map developed by a small group of social work students.

Group mind map on resilience

In Figure 2, the central theme is 'resilience'. The branch words are 'change', incident, regrowth and 'new normal'. The branch words are related to the central theme by demonstrating how one can prepare for future adversity, the role of personal experiences during adversity, healing from an adversity and forming a different lifestyle or creating a 'new normal'. The branch words represent the cycle of building resilience as collaborated by the students. Environmental issues such as climate change have implications for human well-being. Choosing the central topic 'resilience' shows that the students understood sustainable development as a complex process that can have associated challenges on human well-being. Students used the green social work

Figure 2

perspective in the group mind map by using the image of a bird as representing an outside influence such as social, political and cultural forces. Using a green social work perspective involves understanding that outside forces such as systemic barriers or cultural beliefs can influence individuals, families and communities' capacity for growth in a positive or negative way. The students understood social development as shown in the mind maps by reflecting resilience as a positive interaction. The students also showed they understood social development by displaying the word 'support' in their mind map. The group mind map enriched the students learning by working collaboratively and engaging with others in brainstorming sessions.

Case Studies

Case studies were developed by students for students to facilitate participatory learning activities in the classroom. With case-based teaching, students develop skills in analytical thinking and reflective judgment by reading and discussing complex, real-life scenarios (University of Michigan, 2013). The purpose of using a case study in a teaching environment is to present the student with a scenario as close to that which he or she may encounter in practice, in order to work through the problem and devise reasonable and workable solutions.

Students were supported to select and develop their case studies based on the course content and readings, and dedicated time was provided in the classroom. The key components of the case study assignment included the context of the issue/problem, local and global dimensions and implications for social work practice and policy. The case study raises questions to consider in the decision-making process. Students developed case studies and presented them to other students in order to facilitate peer learning. Topics included social work and climate change, disaster (Hurricane Katrina), sustainable communities, indigenous knowledge, human rights and food security. Case studies are known to be an effective teaching tool that involves students more fully in classroom discussion and provides a rich basis for developing students' problem-solving and decision-making skills (Stanford University, 1994). A grading rubric for the case study was developed as shown in Table 1.

Students presented their case studies in class, and were asked to develop at least two discussion questions for their peers. A grading rubric was developed for the class presentation and included the following components: (1) relevance of the case study to social work and sustainable social development, (2) role of case study to create knowledge and insight, (3) critical analysis of context of the problem/local and global dimensions; includes two discussion questions for the class, (4) role of social work in society to address the issue or problem and (5) clear understanding of the topic and issue; verbal skills; use of handouts/PowerPoint/visuals, etc.

Guest Speakers

Practitioners were invited as guest speakers to facilitate student understanding of the potential roles of social work in sustainable social development practice. A variety

Table 1 Grading Rubric for Case Study.

Title page	• APA format is used throughout this document (title page through to references)
Abstract	• Overview of case study; purpose of study; brief description of the context/issue; statement on implication for social work practice/max. 150 words on a single page • Gives readers an overview to see whether they want to keep reading
Introduction	• Description of the context of the issue or problem • Contains things like: the purpose of the case study, the goal of the case study, the aim of the case study, etc. • Relevant topic of importance to social work and sustainable social development
Literature review	• Review of the relevant literature (min. 10 references including peer-reviewed academic journal and materials from the course) • Literature is used to frame the context of the issues or problem; presented in a separate section • The literature review identifies and defines key terms used in the case study
Theoretical framework	• Identify what theories are used to explore the case study (1–2 paragraphs) • A theoretical or conceptual framework in a separate section • For example, green social work, sustainability structural social work, environmental social work, etc.
Local and global dimensions	• Specifically address the local and global dimensions of the problem in the case study. For example, a specific neighbourhood or community, country or region or continent of the world
Implications for social work practice and policy	• Identify several implications for social work practice and policy. This can include what actions are required in social work education, in social work practice and policy, by social workers in their communities countries and internationally
Ethics	• Discuss any important ethical considerations; risks; benefits

of agencies were invited to participate, including community agencies, charitable organizations, environmental groups, advocates and municipal and provincial government ministries. Guest speakers developed PowerPoint presentations on their programs and projects taking place in the local community and province, and resource material was shared in class. The topics presented by guest speakers included community gardens, food security, status of women, emergency management and emergency social services and sustainability and sustainable development at the municipality. Each presentation was followed by a question and answer period in class. The guest speakers shared their practice experience and demonstrated how the concepts of sustainable development and social development are integrated in the field. The presentations highlighted the need to build capacity in social work field education programs to facilitate new placements yet meet [Name of Country Deleted to Maintain the Integrity of the Review Process] Canadian field accreditation requirements for supervision. This is a particular challenge given the need for social work accredited field supervision in Alberta, Canada. The invited guest speakers were provided with a copy of the course outline, and telephone conversations were held prior to their visit in order to answer question and provide additional information. The use of guest speakers is particularly relevant to teaching environmentally focused social work in order to link theory with practice and to foster the development of new social work practice and experiential learning opportunities (e.g. field placements).

Discussion

Social dimensions of environmental problems are elements that need to be further included in the education, mission, tradition, values and practices of modern social work (Peeters, 2012). Social work must assert itself as being essential to addressing the environmental challenges that humanity faces today (Peeters, 2012; Welch, 2012). Upon completion of the course many students expressed that they would have liked to develop an earlier understanding of these issues in their undergraduate education. In the Canadian context there are no required courses on social work and the environment, and students have limited course offerings. This new course is expected to be offered on a bi-annual basis given the demands of required courses in the curriculum. The Canadian accreditation standards do not require core content on the environment. There is a need to build awareness and capacity within the Canadian social work profession, and to integrate new compelling issues outlined in the Global Agenda into accreditation requirements for social work education.

To obtain feedback on student's learning in the course, both formative and summative evaluations were conducted. The completion of the summative evaluation form was voluntary and anonymous, and was offered in addition to the University's Universal Student Rating of Instruction (USRI). Students were asked about their overall experience in the course, whether they have learned new material in the course, and what they liked most and what they found most difficult about the course. Students reported that they enjoyed the interactive learning opportunities, group discussions and non-traditional learning approaches with mind maps and creative activities. Several students highlighted the small class format that enhanced their learning; it is important to note that there were a total of 10 students in the course. Based on course evaluations students provided the following comments:

> I have gained so much knowledge from this course. I have learned about green social work, sustainability, climate change, environmental racism, community gardens, international affairs, [Name of Country Deleted to Maintain the Integrity of the Review Process]'s mining corporations, human rights, international social work and natural disasters just to name a few. The most significant information I have learned in the course was about sustainability and that mitigation work or transformative change is possible! This course taught me that social workers can integrate caring for the environment into daily practice.

The evaluations revealed that some students found the case study assignment challenging in terms of how it differs from a research paper.

There are a number of implications of the use of innovative teaching and learning methods for social work education. Regehr (2013) discusses the need to develop new models of teaching that maximize student experience and learning within a post-secondary environment given trends in higher education. Feedback from students demonstrated that the small class size facilitated interactive learning and participation in the classroom. Students shared their mind maps and case studies in class to facilitate peer learning and to engage in discussion of the key themes, concepts and relationships. Yet, this format may present a challenge for social work education programs facing resource constraints.

There is a pivotal role for social work to play in pursuing a social and environmentally sustainable present and future (Junyent & De Ciurana, 2008; Welch, 2012). Focusing on social and environmental sustainability in social work education, it can contribute to the development of viable ways for humanity to address our current challenges. In addition, there is a strong connection between social and environmental issues and there is tremendous value in making these relationships more apparent and specific in the education and training of social work students (McKinnon, 2008; Welch, 2012). Using a green social work approach the course provided a theoretical understanding of the complexity of developing a holistic and integrated understanding of complex social, economic and environmental phenomena. Education that utilizes collaborative problem-solving activities provides opportunities for students to apply abstract concepts to real world challenges and to link knowledge to action (Dyer et al., 2006; Brundiers et al., 2010). Teaching methods were designed to integrate theory into practice by using individual and group mind maps, case studies, student presentations and guest speakers. Formative and summative evaluations provided feedback throughout the course on student learning.

Conclusion

Sustainability, environmental justice and green social work are of growing interest to social work students, faculty and social workers at local and global levels. Social workers are challenged in their practice to consider complex, interrelated and systemic problems that require new approaches and innovations. As social workers are being called upon to address the impacts of climate change and disasters there is a need to build capacity within the profession to address the realities of environmental degradation at all levels (Drolet & Sampson, 2014; Rebuilding Lives Post Disaster, 2015). This article shares the development and design of a new course on 'social work and sustainable social development' that sought to integrate the social, economic and environmental dimensions into social work practice and policy. A variety of participatory teaching methods were used to facilitate student learning in order to address complex, interrelated and interdependent social, economic and environmental factors.

The course provided students with an in-depth understanding of the complexity inherent in shaping effective responses to social, economic and environmental crises.

Acknowledgments

This work was supported by the Taylor Institute for Teaching and Learning at the University of Calgary, and the Social Sciences and Humanities Research Council of Canada.

Disclosure statement

No potential conflict of interest was reported by the authors.

References

Alston, M., & Whittenbury, K. (2012). *Research, action and policy: Addressing the gendered impacts of climate change*. Dordrecht: Springer Verlag.

Bent-Goodley, T. B. (2015). A call for social work activism. *Social Work, 60*, 101–103. doi:10.1093/sw/swv005

Bochel, C. (n.d.). *Presenting social policy material using mind-maps*. Southampton: The Higher Education Academic Social Work and Social Policy (SWAP).

Boetto, H., & Bell, K. (2015). Environmental sustainability in social work education: An online initiative to encourage global citizenship. *International Social Work, 58*, 448–462. doi:10.1177/0020872815570073

Brundiers, K., & Wiek, A. (2011). Educating students in real-world sustainability research: Vision and implementation. *Innovative Higher Education, 36*, 107–124. doi:10.1007/s10755-010-9161-9

Brundiers, K., Wiek, A., & Redman, C. L. (2010). Real-world learning opportunities in sustainability: From classroom into the real world. *International Journal of Sustainability in Higher Education, 11*, 308–324. doi:10.1108/14676371011077540

Coates, J. (2005). The environmental crisis. *Journal of Progressive Human Services, 16*, 25–49. doi:10.1300/J059v16n01_03

Dale, A., & Newman, L. (2005). Sustainable development, education and literacy. *International Journal of Sustainability in Higher Education, 6*, 351–362. doi:10.1108/14676370510623847

Dominelli, L. (2011). Climate change: Social workers' roles and contributions to policy debates and interventions. *International Journal of Social Welfare, 20*, 430–438. doi:10.1111/j.1468-2397.2011.00795.x

Dominelli, L. (2012). *Green social work: From environmental crises to environmental justice*. Cambridge: Polity Press.

Dominelli, L. (2014). Promoting environmental justice through green social work practice: A key challenge for practitioners and educators. *International Social Work, 57*, 338–345. doi:10.1177/0020872814524968

Drolet, J. (2014). *Social protection and social development: International initiatives*. Dordrecht: Springer.

Drolet, J. (2015). Editorial. *International Social Work, 58*, 351–354.

Drolet, J., & Sampson, T. (2014). Addressing climate change from a social development approach: Small cities and rural communities' adaptation and response to climate change in British Columbia, Canada. *International Social Work*. Advance online publication. doi:10.1177/0020872814539984

Drolet, J., Samuel, M., & Larson, G. (2012). The impacts of natural disasters in diverse communities. *The International Journal of Diversity in Organizations, Communities and Nations, 11*, 95–107.

Drolet, J., Samuel, M., Larson, G., Tesoriero, F. and Ginsberg, L. (January 2010). *International social work and disasters: Our unique and important contribution to disaster relief work and research*. Perspectives, 32, Vancouver: BC Association of Social Workers.

Dyer, A., Selby, D., & Chalkley, B. (2006). A centre for excellence in education for sustainable development. *Journal of Geography in Higher Education, 30*, 307–312. doi:10.1080/03098260600717406

Ersing, R. L., & Kost, K. A. (Eds.). (2012). *Surviving disaster: The role of social network*. Chicago, IL: Lyceum Books.

Global Agenda. (2012). Global agenda for social work and social development. Retrieved from http://www.globalsocialagenda.org/

Hawkins, C. A. (2010). Sustainability, human rights, and environmental justice: Critical connections for contemporary social work. *Critical Social Work, 11*. Retrieved from http://www.uwindsor.ca/criticalsocialwork/theenvironmental-justice-a-critical-connection

Heinonen, T., & Drolet, J. (Eds.) (2012). *International social development: Canadian experience and perspectives*. Winnipeg: Fernwood Publishers.

Jacobs-Lawson, J. M., & Hershey, D. A. (2002). Concept maps as an assessment tool in psychology courses. *Teaching of Psychology, 29*, 25–29. doi:10.1207/S15328023TOP2901_06

Jones, P. (2010). Responding to the ecological crisis: Transformative pathways for social work education. *Journal of Social Work Education, 46*, 67–84. doi:10.5175/JSWE.2010.200800073

Jones, P. (2013). Transforming the curriculum: Social work education and ecological consciousness. In M. Gray, J. Coates, & T. Hetherington (Eds.), *Environmental social work* (pp. 213–230). London: Routledge.

Junyent, M., & De Ciurana, A. M. G. (2008). Education for sustainability in university studies: A model for reorienting the curriculum. *British Educational Research Journal, 34*, 763–782. doi:10.1080/01411920802041343

Kopnina, H., & Meijers, F. (2014). Education for sustainable development: Exploring theoretical and practical challenges. *International Journal of Sustainability in Higher Education, 15*, 188–207. doi:10.1108/IJSHE-07-2012-0059

Lysack, M. (2015). Effective policy influencing and environmental advocacy: Health, climate change, and phasing out coal. *International Social Work, 58*, 435–447. doi:10.1177/0020872814567485

Matthias Barth, M., Godemann, J., Rieckmann, M., & Stoltenberg, U. (2007). Developing key competencies for sustainable development in higher education. *International Journal of Sustainability in Higher Education, 8*, 416–430. doi:10.1108/14676370710823582

McKenzie, S. (2004). *Social sustainability: Towards some definitions* (Hawke Research Institute Working Paper Series No. 27). Retrieved from http://w3.unisa.edu.au/hawkeinstitute/publications/downloads/wp27.pdf

McKinnon, J. (2008). Exploring the nexus between social work and the environment. *Australian Social Work, 61*, 256–268. doi:10.1080/03124070802178275

Noonan, M. (2013). Mind maps: Enhancing midwifery education. *Nurse Education Today, 33*, 847–852. doi:10.1016/j.nedt.2012.02.003

Norton, C. L. (2012). Social work and the environment: An eco-social approach. *International Journal of Social Welfare, 21*, 299–308. doi:10.1111/j.1468-2397.2011.00853.x

Peeters, J. (2012). The place of social work in sustainable development: Towards eco-social practice. *International Journal of Social Welfare, 21*, 287–298. doi:10.1111/j.1468-2397.2011.00856.x

Peterson, A. R., & Snyder, P. J. (1998, August 20–22). *Using mind maps to teach social problems analysis.* Paper presented at the annual meeting of the society for the Study of Social Problems. 48th, San Francisco, CA.

Rebuilding Lives Post Disaster. (2015). Project. Retrieved from www.rlpd.ca

Regehr, C. (2013). Trends in higher education in Canada and implications for social work education. *Social Work Education, 32*, 700–714. doi:10.1080/02615479.2013.785798

Shaw, T. V. (2013). Is social work a green profession? An examination of environmental beliefs. *Journal of Social Work, 13*, 3–29. doi:10.1177/1468017311407555

Stanford University. (1994). Teaching with case studies. *Speaking of Teaching, 5*, 1–4.

Steinemann, A. (2003). Implementing sustainable development through problem-based learning: Pedagogy and practice. *Journal of Professional Issues in Engineering Education and Practice, 129*, 216–224. doi:10.1061/(ASCE)1052-3928(2003)129:4(216)

Trepagnier, B. (2002). Mapping sociological concepts. *Teaching Sociology, 30*, 108–119. doi:10.2307/3211525

University of Michigan. (2013). Case-based teaching and problem-based learning. Retrieved from http://www.crlt.umich.edu/tstrategies/tscbt

UNRISD. (2014). 2010s. Retrieved from http://www.unrisd.org/80256B3C005BB128/%28httpResearchHome%29/$first?opendocument&count=1000

Welch, J. (2012). Sustainability and social development: An integrative examination. *Social Development Issues*, *34*, 56–76.

Zapf, M. K. (2012). Social work and the environment: Understanding people and place. *Critical Social Work*, *11*, 30–46.

Cultivators of Change: Food Justice in Social Work Education

Michelle Lee Kaiser, Sarah Himmelheber, Shari Miller & R. Anna Hayward

Food security is an important social work issue historically, and social work educators are responsible for teaching a curriculum that ensures social workers advance human rights, social justice, and economic justice. Contemporary food justice work focuses on the intersecting issues of policy, health, social justice, economic development, and the natural environment. The long-term global public health and environmental threats posed by the mainstream food system in combination with increasing poverty and food insecurity have led to questions about the ability of communities to sustain a nutritionally adequate and equitably distributed food supply. This paper provides examples of social work courses, units, and assignments that focus on educating students about food and environmental justice issues. Much of this work is based on service learning, which is an effective pedagogical tool for fostering connections between classroom concepts and practice. Courses that help students understand the contextual environments in their local communities provide optimal learning environments to address social, economic, and environmental injustices in the food system. Food justice, in particular, is one lens by which students can learn about environmental justice issues for application to their future practice.

Cultivators of Change: Food Justice in Social Work Education

Globally, the industrialized food system, which relies heavily on non-renewable energy sources, contributes to the pollution of groundwater, soil, and air (McBeath & McBeath, 2009; Story, Hamm, & Wallinga, 2009). Climate change (Lang, 2009), population increases, the loss of arable land (Hoff & Pollack, 1993), and energy demands (Conceição & Mendoza, 2009) have all contributed to rising food prices. These economic and environmental effects have the greatest impact on impoverished households and communities that are more likely to experience food hardships (Conceição & Mendoza, 2009). Contemporary food justice work recognizes both the vulnerabilities and variety of systems involved by focusing on intersecting issues of policy, health, social justice, economic development, and the natural environment.

It is incumbent upon social work educators to nest 'the environment' within the natural world so that social work students frame their interventions from a broader systemic and eco-social perspective. To consider any kind of knowledge outside the context of earth's concerns is to detach everything from its most critical context, to ignore the system of which we are all a part, rendering problem-solving flawed (Orr, 2004). In this case, the problem to be solved by social workers is addressing food justice at multiple system levels. It can sometimes be challenging to see the applied relevance of natural environmental issues for social work. Food issues provide a pragmatic vehicle for clarifying this link. Humans are animals; humans need food to survive; food is a product of the environment and is embedded in the earth's ecosystem; humans who are part of that ecosystem have constructed approaches to producing and accessing food, in the context of the human constructed geo-political, economic landscape, that have contributed to environmental degradation and environmental injustices, as well as multi-systemic issues. These include widespread hunger across the globe (World Food Programme, 2015), chronic diseases and mental health problems related to poor diets (Adams, Gummer-Strawn, and Chavez, 2003; Epstein et al., 2001; Wallinga, Schoonover, & Muller, 2009), unequal access to healthy foods (Apparicio, Cloutier, & Shearmur, 2007; VerPloeg et al., 2009), and unsustainable, exploitative food production methods dependent on limited natural resources (McBeath & McBeath, 2009; Neff, Parker, Kirschenmann, Tinch, & Lawrence, 2011; Polack, Wood, & Bradley, 2008; Worldwatch, 1987, as cited in Hoff & Polack, 1993) and often controlled by multinational corporate industries commodifying food and the environment (Hendrickson, 2003) in the context of neoliberal capitalism.

This paper focuses on how the authors have used food justice as one lens by which social work students can understand the relevance of environmental justice issues in their practice. We provide background information about food justice and its placement within the greater Environmental Principles approach (National People of Color Environmental Leadership Summit Report [National People of Color Environmental Leadership Summit [NPCEL], 1991), connect the issue to US and international teaching standards, discuss pedagogical approaches used, provide extensive examples of how the authors have taught this material at the undergraduate and graduate level, and explain the implications and considerations for incorporating food justice in formal social work education.

Theoretical Framework

Environmental sustainability is a unifying and essential global concern (Scerri, 2009; Schmitz, Stinson, & James, 2010). It is critical for the social work profession to consider how students are being educated around environmental and ecological justice issues, so that they may engage these issues in their future practice. The food system is directly connected to the natural environment in an ecological–social system that has created economic, ecological, and social stressors (Kaiser, 2011; Mullaly, 2006). This ecological–social system, for social workers, should be considered in terms of how humans in this ecological–social system are not only contributors to food insecurity, but also sources of knowledge and resources that can create social change through community food security strategies that emphasize environmental sustainability, public health, cultural appropriateness, self-reliance, and social and economic justice (Kaiser, 2011).

Miller, Hayward, and Shaw (2012) highlight the notion that environmental justice emphasizes that the most vulnerable and marginalized people across the globe are impacted by environmental degradation. This includes people living in food deserts where people lack access to foods that meet their nutritional needs (VerPloeg et al., 2009), farm workers regularly working in fields sprayed with pesticides (Flocks, 2012), and areas where pollution from processing facilities has impacted groundwater (Burkholder et al., 1997; Edwards & Daniel, 1992; Mallin, 2000; Mallin & Cahoon, 2003). Miller et al. (2012) call on the international social work profession to recognize the 'structural power relationships between humans and Earth' (p. 273) and lead the way to utilize an ecological paradigm based on principles of environmental justice.

In 1991, the First National People of Color Environmental Leadership Summit established Principles for Environmental Justice (National People of Color Environmental Leadership Summit [NPCEL], 1991). These principles provide a framework for social work practice in environmental and food justice work. The principles address the importance of 'the interdependence of all species,' an affirmation of ecological unity, and a focus on individual choice in personal and consumer decisions (National People of Color Environmental Leadership Summit [NPCEL], 1991), and provide a framework for social work's focus on issues of food and sustenance in all communities. Individual consumer decision about what is purchased and consumed can be understood as a political activity when one considers food in the context of the interconnection between plants, animals, humans, and the planet. Of course, for many people living in poverty, the lack of choices points to larger local and global systemic issues important to social work. The principles further acknowledge a mandate for ethical and responsible land use and protection of workers (National People of Color Environmental Leadership Summit [NPCEL], 1991). Responsible land use connects with food justice and the need to develop sustainable community food systems, especially in low-income communities (Miller et al., 2012). Similarly, food justice work recognizes that food industry workers, such as those exposed to toxic chemicals in fields or dehumanizing working conditions in the meat industry, should be protected from these health hazards.

Food Justice Pedagogical Framework

Food justice work manifests multiple principles of environmental justice and connects with an array of social work focus areas and practice levels. Due to the number of links, these concepts may be conveyed using different teaching techniques. This section discusses the goodness-of-fit for food justice in social work education, presents pedagogical strategies for its inclusion, and details specific efforts at implementing this content into the social work curriculum.

International Ethical Principles and Standards for Social Work Education

In 2004, the General Assemblies of the International Association of Schools of Social Work (IASSW) and the International Federation of Social Workers (IFSW) adopted the Global Standards for the Education and Training of the Social Work Profession (Sewpaul & Jones, 2004). The IASSW and IFSW state that, 'social work intervenes at the points where people interact with their environments' (Sewpaul & Jones, 2004, p. 2) and that social workers must consider 'interaction among biological, psychological, socio-structural, economic, political, cultural, and spiritual factors' (p.6). The International Federation of Social Workers [IFSW] (2012b) outlines ethical principles, explicitly stating the need to understand how people interact with the natural environment. The IFSW discusses social workers' roles related to globalization and the environment (International Federation of Social Workers [IFSW], 2012a), acknowledging that international trade has resulted in economic inequalities and that limited natural resources should be distributed with consideration for sustainability. These considerations can be understood in the context of neoliberalism and its impacts on the social welfare state and on the globalized, industrialized food system. As such, social workers must understand that poverty, inequality, and environmental degradation are inextricably linked across borders, and threaten the Earth and its inhabitants (International Federation of Social Workers [IFSW], 2012a).

United States Ethical Principles and Standards For Education

Food justice is an important US social work issue that has roots dating back to the Settlement House Movement (Lee, 2011). According to the National Association of Social Workers (2008), social workers are responsible for meeting basic human needs, empowering vulnerable populations, addressing individual and societal well-being, and targeting the contextual environment. Equally important is the call to seek equitable access to resources, support social justice, prevent exploitation, and increase opportunities for oppressed populations (National Association of Social Workers, 2008). In the United States, social work educators are responsible for teaching a curriculum that is directed at ensuring social workers advance human rights, social justice, and economic justice (Council on Social Work Education [CSWE], 2008). Responding to the causes and consequences of food insecurity is a key example of this charge; introducing social work students to food justice work supports their ability to identify oppression and discrimination while responding to the contexts (including the natural environment) that shape practice (Council on Social Work Education [CSWE], 2008).

Pedagogical Strategies

Social work educators can learn from others in higher education that have shown growing interest in food justice issues; teacher-researchers have highlighted connections between food, the social and natural environment, and human well-being (Gottlieb & Fisher, 1996; Sullivan-Catlin, 2002; Wynne, 2006). Service learning and community-based research are two pedagogical strategies for connecting social work education with food justice across the social work curriculum.

Service learning is a pedagogy that integrates the understanding of course material, critical analysis of issues presented, and engagement in the community using different sets of skills (Cress, Collier, & Reitenauer, 2005). For example, Sullivan-Catlin (2002) detailed her efforts to utilize food justice issues as a topical focus within an introduction to sociology course. Through service-learning placements at an urban food bank and a rural emergency food program, students were able to connect their experience with sociological concepts such as stratification, culture, and socialization. Similarly, Wynne (2006) used food as a 'lens ... to engage them [students] in communities of place' (p. 225) by 'help[ing] students see the interconnections between agri-food system restructuring and rural community well-being' (p. 226). Students organized and hosted a Community Food Expo, which resulted in a commitment from campus dining services to increase their procurement of locally sourced foods (Wynne, 2006).

Research methods courses could connect with community-based research opportunities often available through community partners (e.g., food banks, urban farms, advocacy groups). Opportunities for student involvement may include evaluating initiatives aimed at making produce affordable, interviewing residents about food and health, or analyzing data from surveys about food access. In social welfare policy courses, potential activities range from work with local food policy councils to analysis and advocacy related to the US Farm Bill, which is renewed every five years and has implications for conservation, international food prices, and food assistance programs (United States Census, 2014). Critical analysis of challenges related to access, availability, and use of food assistance programs connects well with direct practice methods as well as policy courses. Within micro and macro theory-based courses, food justice provides an opportunity to consider different approaches to the way we understand how humans behave in their environment; these may result in different types of individual and community-based direct services and systems-based solutions to food-related injustices. The following examples explain how this focus on food and the natural environment has been addressed in social work classes, units, and activities at the BSW and MSW levels.

Undergraduate Diversity Course Example

Food was used as a lens to communicate social work ideas throughout an entire diversity course taught at a mid-sized university in the upper-Midwest by Author 2. This service-learning course positioned food as a way to explore dimensions of diversity. Students also contributed two hours per week to their service placements and attended a once-weekly, three–hour seminar. Because of the connection between food justice work and the natural environment, students were able to articulate the importance of environmental considerations in building a culturally appropriate and just food system.

Students self-selected their service placements with one of two Oneida Community Integrated Food System (OCIFS) programs. Developed by the Oneida Tribe, OCIFS aims to address food quality issues, help meet the food needs of tribe members, and create employment opportunities (Oneida Nation, n.d). Students contributed time either to Tsyunhehkwa (joon-henk-wa), loosely translated as 'life sustenance' in Oneida), a working organic farm focused on white corn production and other traditional practices, or Oneida Food Distribution (OFD), a federal commodities program distribution center on the Oneida Reservation. These programs provided students with contrasting experiences, which initially resulted in differing understandings of food justice issues. Students working with OFD focused primarily on their personal experience of providing services to those in need; whereas students working with Tsyunhehkwa increased their awareness of the connection between food access and the environment.

The following excerpt from a student's service-learning journal exemplifies the impact of the Tsyunhehkwa setting:

> At Tsyunhehkwa, they set people up with seed and plants to start their own garden. They are trying to maintain cultural traditions.... The goal of this program is [spreading the idea]... that everyone is capable of having fresh food and food security.

Although OFD did not provide the same opportunities, by having the two groups share their experiences through mid-semester group presentations, students from both service sites were able to make connections between these two OCIFS programs as well as raise critical food justice questions.

Following the OFD group presentation, questions raised included why the tribe was not able to source commodities from its own programs, such as their apple orchards. Discussion after the Tsyunhehkwa presentation centered activities such as seed distribution and the long-term potential of this program to help meet the community's food needs—including community members who received OFD food.

In the classroom, students were introduced to the concepts of intersectionality, privilege, and oppression. In addition to readings from the main text (Anderson & Collins, 2013), each of the five units (race and ethnicity, class, gender, sexuality, and ability) contained at least one reading related to a contemporary food justice issue. By couching concepts such as intersectionality and privilege in food justice content, students were able to logically include the natural environment into their understanding of the human experience.

In the race and ethnicity unit, students investigated the intersection of racial privilege, culture, and land stewardship through a chapter by Lawson (2012) detailing the struggle to save a 13-acre urban farm in Central Los Angeles. In the class unit, students read an essay detailing the personal experience of dumpster diving

(Eighner, 1991) and articles by Clark (2008) and Gross (2012), critiquing food access being the exclusive domain of the marketplace. During the gender unit, assumptions about the relationship between gender and food were challenged through viewing, *Fast Food Women* (Lewis, 1992) and reading articles on women's roles throughout the food production system (Allen & Sachs, 2012; Barndt, 2008). Buerkle (2012) introduced the idea that food choices and eating patterns send messages about power, privilege, and social position in the sexuality unit. In the ability unit, students were exposed to the experience of requiring assistance with eating (Lance, 2012), while Webber, Sobal, and Dollahite's (2012) article reminded students of the centrality of the environment, and that those with different ability statuses and limited resources often face additional food access obstacles.

Food justice was the organizing lens for both the academic and service experiences. Students were exposed to an example of how intersectionality, privilege, and oppression manifest in all systems—including the food system. Attention to the natural environment was strengthened by having one service group involved with sustainable food production and raising the whole group's awareness of the connection between food access, food sovereignty, and environmental stewardship.

Combined BSW/MSW Elective Example

In the context of a critical approach to social work education, Miller and Colleagues created a social work course that invited students to engage with ideas of environmental and ecological justice, with food justice serving as the anchoring point. Food justice and its vicissitudes capture critical aspects of the global impacts of neoliberalism, with the industrialized food system as an elemental example of both inequitable distribution of resources and the unsustainable treatment of natural resources. Not only is this a direct example of global environmental concerns grounded in the geo-political landscape, but it is also captured in the local conditions of the community. This course was offered at a large, public, land-grant university in the Southeastern USA in a community characterized by profound socioeconomic disparity. Approximately 36.7% of people in the county in which the university resides live at or below the US poverty line (United States Census, 2014) and 21.5% are food insecure (Feeding America, 2012). An estimated 77% of children in the county are estimated to be income-eligible for US Federal programs that provide free and reduced breakfast and lunch to students (Feeding America, 2012). Thus, the need for students in this area to develop their critical consciousness and recognize food justice within the larger ecological system is profound.

A transdisciplinary service-learning course was created to benefit the community and provide a learning opportunity for undergraduate and graduate students. A social work educator collaborated with colleagues from the departments of horticulture and foods and nutrition, and the office of service learning to develop an after-school garden program in public elementary (Kindergarten-5th grade) sites in the county. The course was designed iteratively over three semesters and involved a combination of once weekly in-person meetings including students and instructors from the three different disciplines online weekly reflective writing, and three hours per week of service learning at one of five gardens. Students were grouped in cohorts of three (one from each discipline) and were assigned to a site. They collaborated with a group of children (K-5th) to grow and maintain a food garden on their school's grounds. The university students facilitated programs and learning activities with the children, designed to enhance the children's understanding of their relationship to food, where food comes from, their sense of self-efficacy, empowerment, and social growth.

The university students simultaneously were working on building their own knowledge about food systems and food justice in the context of ecological justice, their knowledge about the community and about the family–school–community relationships, and their understanding of their respective and combined roles in this learning experience. The course relied heavily on collaboration—with the community, across disciplines, and between and among students and faculty. Students responded to weekly reflective journal prompts throughout the semester; based on student reflections in the aggregate a few clear themes emerged: social work students struggled to clarify their roles on the transdisciplinary team at the beginning of the semester, and by the end of the semester recognized their roles as facilitators of communication, relationship building, and making linkages between and across systems. Students began the semester with a vague understanding of the links between social work and the natural environment, and where food 'fits,' and by the end of the semester often reflected on how clear the links became through the combination of multi-disciplinary readings, discussions, and applied experiences in the school-community gardens with the children. Students often noted how overwhelmed they were to learn the true scope of food insecurity issues for the children (and families) with whom they worked. Students indicated that participation in this learning experience changed their relationship to social work and thoughts about how they plan to engage in practice going forward.

MSW Advanced Practice Elective Example

Food justice issues can also be addressed within courses that focus more broadly on issues of environmental and ecological justice. For example, Author 4 developed an advanced practice elective for master's level students that focused on environmental justice issues including land, water, food, and the disproportionate impact of environmental hazards on economically disadvantaged communities. Although food justice issues were woven throughout the course, a specific unit focused on food, community agriculture, and sustainable communities.

The unit began with a discussion of food insecurity, poverty, and the idea of urban food deserts, a concept that was new to most of the suburban middle-class students who enrolled in the class. Guest speakers from the university's department of nutrition and the sustainable studies program, along with local farmers, provided students with a broader perspective on the location and availability of fresh, local produce. Students also researched the relationship between food choices and other areas of life, including mental and physical wellness, school performance, and mental health treatment.

Students each contributed six service-learning hours during the semester; many students completed these hours at a community garden where the class as a group provided volunteer service.

The food unit was the most dynamic for the students and issues related to food justice reoccurred throughout the semester when the focus shifted to land use, conservation, climate change, and water issues. Many students were unaware of local food justice issues and community garden efforts. Including faculty from other departments broadened students' focus beyond traditional social work areas. Weekly assignments charged students with the task of reflecting on research, position papers, or their service experiences. Students indicated that learning about food justice issues including food production and food insecurity had the most immediate impact on their worldview. In reflection papers and more formal course evaluations, they shared that being introduced to these issues resulted in personal behavior changes (diet modifications were noted by several students) and that course content has helped clarify the connections between the broader natural environment, their own personal choices, and the choices and opportunities of their clients and local communities. Issues of food justice provided the most concrete and accessible examples of environmental justice issues for students. Connecting with both urban (rooftop garden) and rural (large-scale farmers) guest speakers who shared similar concerns about environment and food justice helped to bridge gaps in their understanding of the relevance of these issues for social work.

Food Justice Activity Example

Kaiser created an activity used in undergraduate and graduate classes, residence halls, continuing education classes, and community groups. This activity is akin to a poverty simulation (Speakman, n.d.); however, it integrates real-life scenarios the author experienced working in Central Appalachia as well as through interviews with hundreds of food pantry clients in a Midwest state. The hour-long experience and 30-minute discussion addressed four key areas of food justice: societal/economic conditions (i.e., manufacturing plant layoffs due to companies moving outside of the USA, challenges related to health care access, health conditions associated with working in extraction industries), accessibility of food (i.e., transportation, distance to food sources), availability of food (i.e., different types of food sources available, challenges of finding specific foods prescribed for chronic diet-related health conditions), and affordability of food (i.e., actual costs of different types of food, tradeoffs with other household budget items). Small groups of participants worked their way through real-life scenarios of families as they make decisions around food. At three points in the activity, participants choose between two options, which then directs the way the story unfolds.

The food security simulation highlights the relationships among people, food, and the natural environment by specifically exploring the impact of environmental degradation due to logging that created costly housing damage and the reality of the health risks experienced by persons working in this industry for a family challenged by poverty.

Additional issues related to why foods such as fruits and vegetables are more expensive than foods laden with high fats and high fructose corn syrup invited discussions about impacts of US food policy. A 2014 participant noted on the evaluation that the simulation 'effectively integrates the many complex issues including environmental, social, and community health involved with food production and hunger.'

Implications for Social Work Education

Incorporating environmental and ecological justice issues in the social work curriculum benefits students, faculty, schools of social work, colleges and universities, and the communities within which this work occurs. An emphasis on food systems and food justice makes salient both critical need and the scope of potential benefit. When this type of content is included, social work students move past the construct of the environment as purely social; they begin to recognize an environment that is much more encompassing and grounded in the necessity of increased local and global understanding. This shift in perspective not only enhances students' critical consciousness, but also their content knowledge and practice skill development. This has implications for how they will practice as social workers and how they will participate in the world around them as humans. When employing community-based approaches to education, faculty can build rich, collaborative relationships across disciplines, chipping away at the artificial divides that may impede education and knowledge building. These approaches promote participatory and collaborative town–gown relationships. When university faculty, staff, and students come together to join a community with its members and resources, the potential to directly address far-reaching, persistent, and emerging social concerns is profound.

While many benefits exist for students, communities, educational institutions, and faculty, incorporating environmental justice issues into the curriculum does present challenges. The authors have found that part of the process of addressing environmental justice in social work classes is the need to educate other colleagues about the intersections of environmental issues across different practice levels and across the core curriculum. Food justice offers an appealing entry point for multiple social work content areas. Faculty may be required to teach certain courses or have limited flexibility.

Faculty should consider the number of students needed to make a new course financially viable to their college or departmental unit. Options may include first developing those courses as an independent study, group studies, or special topics course. The course may then be tested a few times with rigorous evaluation. Faculty should take advantage of their institutional resources. Many institutions offer service-learning mentorship or course design assistance. Others may offer grants to develop courses, with money used for everything from transportation to honoraria for community partners.

Lastly, social work faculty should consider working with student organizations interested in food and environmental justice issues. Faculty should develop their community-based partnerships and encourage working with those partners to develop service-learning experiences related to social work direct practice, advocacy, and policy work. Kaiser supervises a student group who has worked with a children's hunger program to develop a school breakfast scorecard, conducted focus groups about food and the environment with residents with an urban farm, assisted in the evaluation of a healthy corner store initiative, and worked with the health department to develop Web-based local food guides. Author 2 advises a student organization focused on reducing food waste and redirected unused, nutritious food to vulnerable community members in need of meals. These experiences further engage students in their own learning about the interconnectedness of health, social justice, economic development, and the natural environment.

Conclusion

It is crucial that social work educators recognize the role they have in teaching students about the relevance of environmental justice to social work practice. Food justice is one lens by which students can understand the interconnectedness of food, people, and the natural environment. As the global population continues to increase and contemporary industrialized global food production seeks to meet the increased demand for food through unsustainable practices that deplete natural resources, the concern for continued widespread hunger and pollution from food production, processing, distribution, will impact the health and livelihood of individuals across the world, especially those who are the most vulnerable. Food justice work is environmental justice work. Social workers must learn about the historical, political, cultural, and ecological causes and consequences related to food and the environment in order to best serve individuals, households, and communities. The inclusion of different examples across the undergraduate and graduate curriculum provides tangible opportunities for educators to incorporate these universal issues of food justice and its relationship to environmental principles. It provides a starting point for educators to consider these applications in core areas such as research, practice, and policy. Social work students must engage in this material to further develop themselves professionally and as citizens of the world.

Disclosure statement

No potential conflict of interest was reported by the authors.

References

Adams, E. J., Grummer-Strawn, L., & Chavez, G. (2003). Food insecurity is associated with increased risk of obesity in California women. *Journal of Nutrition, 133*, 1070–1074.

Alaimo, K., Olson, C. M., & Frongillo, Jr, E. A. (2001). Food insufficiency and American school-aged children's cognitive, academic, and psychosocial development. *Pediatrics, 108*, 44–53.

Allen, P., & Sachs, C. (2012). *Women and food chains: The gendered politics of food.* In P. Williams-Forson & C. Counihan (Eds.), *Taking food public: Redefining foodways in a changing world* (pp. 23–40). New York: Routledge.

Anderson, M. L., & Collins, P. H. (2013). *Race, class, and gender: An anthology* (8th ed.). Belmont, CA: Wadsworth.

Apparacio, P., Cloutier, M-S., & Shearmur, R. (2007). The case of Montreal's missing food deserts: Evaluation of accessibility to food supermarkets. *International Journal of Health Geographics, 6*(4), 1–13. doi:10.1186/1476-072X-6-4

Barndt, D. (2008). Whose 'choice'?: 'Flexible' women workers in the tomato food chain. In C. Counihan & P. Van Esterik (Eds.), *Food and culture* (pp. 452–466). New York: Routledge.

Buerkle, C. W. (2012). Metrosexuality can stuff it: Beef consumption as (heteromasculine) fortification. In P. Williams-Forson & C. Counihan (Eds.), *Taking food public: Redefining foodways in a changing world* (pp. 251–264). New York: Routledge.

Burkholder, J. M, Mallin, M. A., Glasgow, Jr, H. B., Larsen, L. M., McIver, M. R., Shank, G. C., & (1997). Impacts to a coastal river and estuary from rupture of a large swine waste holding lagoon. *Journal of Environment Quality, 26*, 1451–1466. doi:10.2134/jeq1997.00472425002600060003x

Clark, D. (2008). The raw and the rotten: Punk cuisine. In C. Counihan & P. Van Esterik (Eds.), *Food and culture* (pp. 411–422). New York: Routledge.

Conceição, P., & Mendoza, R. U. (2009). Anatomy of the global food crisis. *Third World Quarterly, 30*, 1159–1182. doi:10.1080/01436590903037473

Council on Social Work Education (CSWE). (2008). Educational policy and accreditation standards. Retrieved from http://www.cswe.org/File.aspx?id=13780

Cress, C. M., Collier, P. J., & Reienauer, V. L. (2005). *Learning through serving: A student guidebook for service-learning across the disciplines*. Sterling, VA: Stylus Publishing.

Edwards, D. R., & Daniel, T. C. (1992). Environmental impacts of on-farm poultry waste disposal: A review. *Bioresource Technology, 41*, 9–33. doi:10.1016/0960-8524(92)90094-E

Eighner, L. (1991). On dumpster diving. *Three Penny Review, 47*, 6–8.

Feeding America. (2012). Data by county in each state. Retrieved from http://www.feedingamerica.org/hunger-in-america/our-research/map-the-meal-gap/data-by-county-in-each-state.html

Flocks, J. D. (2012). The environmental and social injustice of farmworker pesticide exposure. *Georgetown Journal on Poverty Law & Policy, 19*, 255–282.

Gottlieb, R., & Fisher, A. (1996). *Community food security and environmental justice: Searching for a common discourse. Agriculture and Human Values, 3*, 23–32.

Gross, J. (2012). Capitalism and its discontents: Back-to-the-lander and freegan foodways in rural Oregon. In P. Williams-Forson & C. Counihan (Eds.), *Taking food public: Redefining foodways in a changing world* (pp. 71–87). New York: Routledge.

Hendrickson, M. (2003). An overview of concentration in the food system. Retrieved from http://www.foodcircles.missouri.edu/leopold/sld001.htm

International Federation of Social Workers [IFSW]. (2012a). Globalisation and the environment. Retrieved from http://ifsw.org/policies/globalisation-and-the-environment/.

International Federation of Social Workers [IFSW]. (2012b). Statement of Ethical Principles. Retrieved from http://ifsw.org/policies/statement-of-ethical-principles

Kaiser, M. L. (2011). Food security: An ecological-social analysis to promote social development. *Journal of Community Practice, 19*, 62–79. doi:10.1080/10705422.2011.550261

Lance, G. D. (2012). Do the hands that feed us hold us back? Implications of assisted eating. In P. Williams-Forson & C. Counihan (Eds.), *Taking food public: Redefining foodways in a changing world* (pp. 299–305). New York: Routledge.

Lang, T. (2009). Reshaping the food system for ecological public health. *Journal of Hunger & Environmental Nutrition, 4*, 315–335. doi:10.1080/19320240903321227

Lawson, L. (2012). Cultural geographies in practice: The south central farm: Dilemmas in practicing the public. In P. Williams-Forson & C. Counihan (Eds.), *Taking food public: Redefining foodways in a changing world* (pp. 89–92). New York: Routledge.

Lee, L. Y. (2011). Hungry for peace: Jane Addams and the Hull-House Museum's contemporary struggle for food justice. *Peace & Change, 36*, 62–79. doi:10.1111/j.1468-0130.2010.00671.x

Lewis, A. (Director) (1992). *Fast food women (motion picture)*. Whitesburg: Appalshop Film and Video.

Mallin, M. A. (2000). Impacts of industrial animal production on rivers and estuaries. *American Scientist, 88*, 26–37. doi:10.1511/2000.1.26

Mallin, M. A., & Cahoon, L. B. (2003). Industrialized animal production: A major source of nutrient and microbial pollution to aquatic ecosystems. *Population and Environment, 24*, 369–385. doi:10.1023/A:1023690824045

McBeath, J., & McBeath, J. H. (2009). Environmental stressors and food security in China. *Journal of Chinese Political Science, 14*, 49–80. doi:10.1007/s11366-008-9036-4. doi:10.1007/s11366-008-9036-4

Miller, S. E., Hayward, R. A., & Shaw, T. V. (2012). Environmental shifts for social work: A principles approach. *International Journal of Social Welfare, 21*, 270–277. doi:10.1111/j.1468-2397.2011.00848.x

Mullaly, B. (2006). *The new structural social work: Ideology, theory, and practice* (3rd ed.). Toronto: Oxford University Press.

National Association of Social Workers. (2008). *Code of ethics*. Washington, DC: NASW.

National People of Color Environmental Leadership Summit [NPCEL]. (1991). *Principles of environmental justice*. New York: United Church of Christ Commission for Racial Justice.

Neff, R. A., Parker, C. L., Kirschenmann, F. L., Tinch, J., & Lawrence, R. S. (2011). Peak oil, food systems, and public health. *American Journal of Public Health, 101*, 1587–1597. doi:10.2105/AJPH.2011.300123

Oneida Nation. (n.d.). Oneida community integrated food system. Retrieved from http://www.oneidanation.org/ocifs/.

Orr, D. W. (2004). *Earth in mind: On education, environment, and the human prospect*. Washington, DC: Island Press.

Polack, R., Wood, S., & Bradley, E. (2008). Fossil fuels and food security: Analysis and recommendations for community organizers. *Journal of Community Practice, 16*, 359–375. doi:10.1080/10705420802255114

Scerri, A. (2009). Paradoxes of increased individuation and public awareness of environmental issues. *Environmental Politics, 18*, 467–485. doi:10.1080/09644010903007344

Schmitz, C. L., Stinson, C. J., & James, C. D. (2010). Community and environmental sustainability: Collaboration and interdisciplinary education. *Critical Social Work, 11*, 83–100.

Sewpaul, V., & Jones, D. (2004). Global standards for the education and training of the social work profession. *Adopted at the General Assemblies of the International Association of Schools of Social Work (IASSW) and the International Federation of Social Workers (IFSW) in Adelaide, Australia*. Retrieved from http://cdn.ifsw.org/assets/ifsw_65044-3.pdf

Speakman, D. (n.d.). Ohio Association of Foodbanks. *Poverty simulation: An interactive experience aimed at helping the community understand poverty*. Retrieved from http://www.ohiofoodbanks.org/docs/programs/povertysimulation.pdf

Story, M., Hamm, M. W., & Wallinga, D. (2009). Food systems and public health: Linkages to achieve healthier diets and healthier communities. *Journal of Hunger & Environmental Nutrition, 4*, 219–224. doi:10.1080/19320240903351463

Sullivan-Catlin, H. (2002). Food, hunger, and poverty: A thematic approach to integrating service learning. *Teaching Sociology, 30*, 39–52. doi:10.2307/3211519

United States Census. (2014). State and county quick facts. Retrieved from http://quickfacts.census.gov/qfd/states/.

United States Department of Agriculture [USDA]. (2014). 2014 Farm Bill highlights. Retrieved from http://www.usda.gov/documents/usda-2014-farm-bill-highlights.pdf

VerPloeg, M., Breneman, V., Farrigan, T., Hamrick, K., Hopkins, D., Kaufman, P., & Tuckermanty, E. (2009). Access to affordable and nutritious food-measuring food deserts and their consequences: Report to congress. *USDA ERS Administrative Publication No. (AP-036).*

Wallinga, D., Schoonover, H., & Muller, M. (2009). Considering the contribution of US agricultural policy to the obesity epidemic: Overview and opportunities. *Journal of Hunger & Environmental Nutrition, 4*, 3–19. doi:10.1080/19320240802706817

Webber, C. B., Sobal, J., & Dollahite, J. S. (2012). Physical disabilities and food access among limited resource households. In P. Williams-Forson & C. Counihan (Eds.), *Taking food public: Redefining foodways in a changing world* (pp. 127–137). New York: Routledge.

Williams-Forson, P. (2008). More than just the 'big piece of chicken'. The power of race, class, and food in American consciousness. In C. Counihan & P. Van Esterik (Eds.), *Food and culture* (pp. 342–353). New York: Routledge.

World Food Programme. (2015). World food programme: Fighting hunger worldwide. *Hunger.* Retrieved from http://www.wfp.org/hunger

Wynne, W. D. (2006). Civic engagement through civic agriculture: Using food to link classroom and community. *Teaching Sociology, 34*, 224–235. doi:10.1177/0092055X0603400302

Exploring the Intersection of Environmental Degradation and Poverty: Environmental Injustice in Nairobi, Kenya

Jennifer L. Willett

Environmental justice explores the nexus between structural inequalities and environmental degradation. Although scholarship on environmental justice is vast, this literature is centred in developed countries, rather than exploring the injustices occurring in international contexts. This study aims to address this gap through a combined phenomenological and ethnographic approach in Jam City, a poor community in Nairobi, Kenya. Findings that are discussed include the similarities of environmental injustices between Jam City and the established literature as well as several nuances with the literature, including how lack of resources for response perpetuates injustice and how environmental hazards may not be explicitly dangerous but can still cause disparate harms. This study supports the argument that social workers should be more involved in promoting environmental justice by increasing our focus on the topic in our curriculum, research and practice.

Introduction

Environmental problems mirror oppressive structural forces in society, as the primary victims of environmental degradation are also the main victims of injustices related to class, gender, race, ethnicity and locale (Nixon, 2011). These groups, particularly the poor in developing countries, bear a disproportionate burden of environmental degradation. However, the conceptual frame of environmental justice has not been widely used to explore international contexts but tends to focus on environmental hazards in the Global North. This study begins to fill this gap by assessing the applicability of environmental justice in Jam City, a poor community in Nairobi,

Kenya. This study also provides a case of international environmental injustice to use as a teaching tool regarding these issues.

Literature Review

Environmental justice is at once a social activism movement, a policy discourse and a conceptual frame (Agyeman, 2014; Kütting, 2004) but the values underpinning the cause remain steady across its uses. Environmental justice argues that environmental degradation and power imbalances are joined (Martinez-Alier, 2014; Nixon, 2011). Structural conditions drive hazardous conditions to the most marginalized people (Robbins, 2014). Thus, the main victims of environmental injustice are also typically victimized by other injustices (Mohai, Pellow, & Roberts, 2009; Nixon, 2011).

A key component of this argument is that decisions about environmental harms are rooted in oppression and control (Sikor & Newell, 2014). Therefore, there is little participation from marginalized communities in decisions about the environmental risks and hazards that affect them (Sikor & Newell, 2014). These communities also often do not have equal access to the benefits that result from the environmental changes, which perpetuates the injustice of disproportionate environmental burdens (Minkler, Vasquez, Tajik, & Petersen, 2008). The importance of this intersectionality of marginalization and environmental degradation is because of the disparate impacts in wellbeing and quality of life in affected communities (Ravi Rajan, 2014). Thus, the demands for the movement are simple: environmental fairness and structural equality (Robbins, 2014).

The environmental justice movement translates to a conceptual framework from which to understand the relationships among disparate environmental benefits and harms, and structural inequalities. Thus, when using environmental justice as a framework scholars focus on disproportionate environmental degradation in marginalized communities, the related unjust impacts, and/or the lack of participation in environmental decisions, and then situate these scenarios within the socio-political context. The empirical literature that utilizes this framework is vast and includes hundreds of studies (Mohai et al., 2009; Robbins, 2014). A portion of these were headed by environmental social workers (Coates & McKay, 1995; Peeters, 2012; Rogge, 1994; Silver, 1994; Soine, 1987). The importance of environmental justice continues to spread throughout the profession and is now included within the 2015 Educational Policy and Accreditation Standards for social work in the US and is considered a core issue within the international social work subfield (Council on Social Work Education (CSWE), 2014; IFSW, IASSW, & ICSW, 2012). However, despite the extensive scholarship and the recognition of its place within the profession, gaps in understanding remain.

Although environmentalists have variably connected environmentalism and social justice throughout the modern environmental movement, most scholars agree that the concept of environmental justice blossomed out of the Civil Rights and labour struggles in the US (Mohai et al., 2009; Ravi Rajan, 2014; Stretesky, Huss, Lynch, Zahran, & Childs, 2011; Strong & Hobbs, 2002). The tipping point for recognition of environmental justice, then termed environmental racism, occurred in North Carolina between 1978 and 1982. Following the illegal dumping of over 30,000 gallons of oil tainted with polychlorinated biphenyl (PCB) (a known carcinogen) along the highway, the state government built a landfill to house PCBs in the predominately black and poor Warren County (Strong & Hobbs, 2002). The landfill was built without community input and against community wishes, which resulted in widespread protests and mass arrests (Strong & Hobbs, 2002). The case of Warren County is often cited as the first documented case of environmental injustice, and it led to several official reports and legal precedents about environmental justice (Strong & Hobbs, 2002).

Much of the literature has remained along similar the lines of inquiry. Studies tend to be racialized (Agyeman, 2014; Sikor & Newell, 2014), which has a high applicability to communities in the US but may not apply in other countries. Studies also tend to focus on the intentional placement of hazardous sites in marginalized communities, which does not include disparities within the environment that develop without racist, unjust, or malicious intent (Mohai et al., 2009; Robbins, 2014).

Further, despite documentation that the environmental degradation and lack of power of the victims of environmental injustice tends to be a dangerous 'threat multiplier' for the worldwide poor (Nixon, 2011, p. 3), environmental justice scholarship largely remains within the confines of the Global North (Sikor & Newell, 2014). Indeed, many scenarios that could have been explored through environmental justice used other conceptual frames like the human rights to control resources, food and water; disasters; and participation (Dominelli, 2012; Maathai, 2009; Ravi Rajan, 2014; Saro-Wiwa, 1995; Shiva, 2005). Thus, the international literature on environmental justice is scant and questions remain about the applicability of environmental justice in developing counties. However, utilizing the framework of environmental justice is ideal in these scenarios as it equally highlights the environmental degradation in marginalized communities and the related structural forces, along with the disparate impacts that occur as a result of these problems.

Among the international studies that utilize environmental justice as a frame, there is a clear focus on the inequities of global processes. These include shifting the burden of e-waste to the Global South, unfair resource extraction in the Global South by transnational corporations, and the injustice of climate change (Mohai et al., 2009; Sikor & Newell, 2014; Willett, 2015). This trend has been cautioned against because the unique structural conditions in marginalized communities tend to lead to uneven environmental injustice across communities (Kütting, 2004). Instead, scholars and activists are encouraged to explore environmental justice as an intra-society, locally contextualized issue.

Scholars have identified these gaps in understanding and have made a call for more application of the environmental justice framework to international contexts (Nixon, 2011; Robbins, 2014; Sikor & Newell, 2014; Willett, 2015). This is an important expansion because it could connect impacted communities with activists and one another around the world (Robbins, 2014). Further, by exploring environmental justice in other communities, the most marginalized people have their voices heard, rather than being incorporated into the experiences of other communities (Nixon, 2011).

This study aims to fill these gaps by exploring environmental justice in Jam City, a poor neighbourhood in Nairobi, Kenya. In addition, this study provides a case of environmental injustice that can be used to support student learning in these areas.

Context

There are numerous forces at play in the Kenyan context and poor Kenyans are acutely affected by all of them, many of which are out of their control and rooted in historical processes (Maathai, 2009). Kenya still struggles to overcome the legacy of 60 years of British colonialism (Hornsby, 2013). The first free Kenyan leader, Jomo Kenyatta, was supported by the British and he kept the established neoliberal system in tact (Hornsby, 2013). The leaders that have followed Kenyatta espoused the same philosophy. Thus, the economic system of Kenyan elevates a few, divides and oppresses the many, and degrades the environment (Wa Kinyatti, 2010).

As a result, many Kenyan communities continue to have poor living standards and low development. Income inequality and poverty have consistently grown since independence (Ndege, 2009). In fact, living standards in an independent Kenya are now lower than living standards in a colonized Kenya (Wrong, 2009). Nearly half of Kenya's 40 million citizens live in poverty (UNDP, n.d.). Life expectancy is just 62 years (UNDP, n.d.). The Kenyan government is limited and is not able to provide sufficient services for the masses in need (Watsa & Warutere, 2013). There are also not enough social work educational programs in the country nor are there sufficient trained social workers on the ground with the resources needed to support the poor (Mungai, Wairire, & Rush, 2014).

Relatedly, Kenya has one of the highest population growth rates in the world. Between 1999 and 2008, the number of Kenyans grew from 28.8 million people to 38 million people, an increase of 33 percent (National Coordinating Agency for Population and Development (NCAPD), 2010). As the population expanded, Kenya has urbanized into unplanned and unsupported slums in Nairobi (Population Action International (PAI), 2012). It is estimated that 60 percent of Nairobians live in poverty in one of the many Nairobi slums (Oxfam GB Kenya Programme, 2009). Living conditions in slums are harsh. Residents must contend with many difficulties that are not as prevalent in the wealthy areas like environmental degradation and hazards, higher HIV rates, higher child mortality rates, poor support networks, higher crime rates and higher relative costs related to poor governance (bribing officials) (Oxfam GB Kenya Programme, 2009). The economy in Kenya also does not include enough jobs for the present population and unemployment is expected to grow along with the population (National Coordinating Agency for Population and Development (NCAPD), 2010). Therefore, these urban problems will likely continue to affect poor Kenyans.

The environment in Kenya is fragile and has been abused for generations. Kenya was originally colonized to support the Industrial Revolution of Great Britain through the provision of new sources of labour, resources and markets (Maathai, 2009; Wa Kinyatti, 2010). The environment degraded quickly due to the insatiable British quest for natural resources and the implementation of an extensive export-focused farming system (Hornsby, 2013; Ndege, 2009). The resulting environmental degradation has not been repaired. The environment continues to be damaged in other ways in urban areas. Public spaces are under threat from developers (Makworo & Mireri, 2011). Nearly half of the urban population does not have access to clean water (WaterAid, n.d.). Three quarters of Kenyan citizens do not have access to functioning solid waste management systems (WaterAid, n.d.). Air pollution in the form of high particulate matter levels from unregulated vehicle emissions is dangerous (Kinney et al., 2011). Poor Kenyans are at particular risk of harm from these environmental problems.

Exploring the impacts of environmental injustice in the slums of Nairobi is an analytically productive case in the intersecting forces of structural inequalities and degraded environments that combine to form environmental injustice. This study will also add voices to the environmental justice literature of people who are affected, adding needed scholarship that is based in a developing country. In addition, this case will concretize amorphous environmental injustice issues, which will provide a tool for learning about these issues.

Methods

Advisory Board

An Advisory Board of people in the study population, poor Kenyans who were affected by environmental problems, guided this study. By collaborating with a local Advisory Board, this study encourages a local viewpoint, insures the inclusion of topics important to the affected populations, and promotes the trustworthiness of the data (Corbin & Strauss, 2008; Padgett, 2008). In addition, by using an Advisory Board this study supports the human right of participation in the research project itself (Sepulveda Carmona, 2013). The Advisory Board members were invaluable to the success of this study through their introductions to the local communities and their explicit support for the study, which made participants more comfortable speaking with the researcher, a white American woman.

All of the Advisory Board members were young men. Like most of the participants, all Advisory Board members were part of the Kamba ethnic group, the fifth largest ethnic group in Kenya (Hornsby, 2013). Although the Kamba often have political power in the role of a swing vote in presidential elections, the Kamba endure social exclusion during non-election times as other ethnic groups form the ruling party and the dissent. Due to their ethnicity, the Advisory Board members lived through marginalization on a daily basis.

Data Collection and Sample

Data were collected through a combined phenomenological and ethnographic approach to explore the shared meanings that underlie the phenomena of study (environmental injustice), the stories that people speak of regarding the phenomena, and the structural and contextual conditions (Padgett, 2008). Data included field notes, ethnographic interactions and observations, and formal interviews. The flexibility of the data collection adhered to phenomenological and ethnographic standards (Padgett, 2008).

Formal interviews

Formal interviews were conducted using a semi-structured interview guide that was revised throughout the study based on Advisory Board suggestions. Questions focused on environmental degradation in participant communities. The interviews typically lasted for approximately one hour.

Participants for the formal interviews were gathered through a snowball sample that was guided by the Advisory Board. The participants were often transient across Nairobi neighbourhoods and between Nairobi and their home communities. Thus, participants were not asked to represent one geographical area but rather discuss current and former environmental problems in any areas which they considered part of their community. Inclusion criteria for participation in the formal interviews was poverty and first-hand experience with an environmental problem. In total, 32 participants were formally interviewed, which is a comparatively large sample size for the phenomenological approach (Padgett, 2008). This larger phenomenological sample size allowed for data that is both deep and broad (Baker & Edwards, 2012).

Cultural norms often shaped the context of the interviews. For example, many participants felt more comfortable discussing their communities to an outsider in a group. Even so, the formal interviews were challenging and indicated participant discomfort with the formal interview process.

Ethnographic interactions and observations

When participants were asked to show the devastation of the environmental problem, their stories came alive. Thus, in most cases, the interview was a starting point to the data collection but the community tours were when rich data were collected. These community tours lasted between a few hours to over the course of several days.

Often during these tours, community members who were not formal interviewees wanted to discuss their interest in the project. These were recorded as ethnographic interactions. In other instances, ethnographic interactions and observations related to the project occurred informally by virtue of the researcher living in Nairobi. In addition to providing rich data, the ethnographic observations and interactions allowed for the consideration of processes of which the participants were not consciously aware and did not speak about (Corbin & Strauss, 2008; Padgett, 2008). Including these observations and interactions contributes to a more holistic understanding of participant experience and enhances rigor (Corbin & Strauss, 2008). The 'sample' for ethnographic interactions and observations was variable beyond the inclusion criteria and also included wealthy Kenyans, NGO workers and people from all ethnic groups.

Most informal interactions and formal interviews were conducted in English or Swahili, both of which are spoken by the researcher. In a few cases participants spoke Kikamba, which was translated by an Advisory Board member. Data collection occurred over six months as part of a larger study on the impacts of environmental degradation on poor communities in Kenya.

Data Analysis

The data were explored through three waves of analysis using a phenomenological approach in NVivo for Mac. In this approach, the essence of the shared experience around the identified phenomenon of environmental injustice was the focus of analysis (Padgett, 2008), rather than a number of other shared experiences. The findings provide a 'thick description' of a case of environmental injustice in Kenya, in order for outsider readers to understand what it is like to experience environmental injustice as the participants did (Padgett, 2008).

In the first wave of analysis, the data were inductively open coded to develop a codebook (Corbin & Strauss, 2008). In the second wave of analysis, the data were re-coded to check interpretation and ensure all of the data was coded in the same way. In the third wave of analysis, horizontal and comparative analyses were conducted to explore the linkages across the data, and the similarities and differences across themes and communities (Corbin & Strauss, 2008). To support validity, the Advisory Board member-checked all findings.

The Case of Flash Flooding in Jam City

Jam City is a formerly rural area in Machakos County, about 19 miles outside of Nairobi, situated at a bend in the Athi River. It is an ideal location: just off the highway, near Nairobi and the airport, but outside of the immediate bounds of the city. The physical community itself was organized and small; it was just two rows of small iron-sheet structures that housed people and businesses. However, the amount of people living in the community was large, about 600 adults and 1,400 children, as reported by the village chief.

Jam City was visibly poor. The houses were constructed primarily of corrugated iron sheeting panels. The 'elite' in the community, like the village chief, owned rudimentary concrete-block houses. The residents in this community lived on this land for many generations and farmed the nearby vacant land. The residents referred to Jam City as their village, a term that is typically used for rural communities. And indeed, the community felt like a small tight-knit village. The residents knew each other by name and role (i.e. that's Mary; she escorts the children to the school across the highway) where in other communities participants did not know most of their neighbours.

Nairobi was rapidly sprawling outward and its boundaries had begun to engulf Jam City, greatly changing the community. Perhaps the most detrimental change came in 2011 when a Chinese development company built a luxurious gated apartment complex just three inches from the edge of Jam City for Nairobi commuters. It was a shocking site of inequality. On the inside of the development's gate, there were several high-end apartment towers with manicured gardens, guard stations and a children's playground; around Nairobi, there were several billboards advertising this development and its luxury. On the outside of the gate, was Jam City with no electricity, running water, adequate food supply, or consistent source of employment.

Environmental Degradation and Secondary Effects

Although not identified in such specific terms, the environmental injustice that stemmed from this development was the main topic of the research conducted in Jam City. The participants explained that the new development was built on a natural flood plain at a bend in the Athi River. Before the development, this area flooded during the rainy season and the water subsided naturally. Jam City was located just outside of this flood plain purposefully. Because of the decision to build apartments in the flood plain, the developers built a floodwall around the complex to protect the property. Due to the wall, all floodwaters were inadvertently directed into Jam City, which flowed downhill through the entire village until again meeting with the river during each rain.

The floods created an emergency situation for the residents of Jam City. Participants collaboratively explained in a group interview:

> It comes at night ... If you find a flood, you go around and knock on the houses. If there are small children or the elderly people, we have to carry them and support them ... We help them walk out ... We just go near the main road. Yeah, we stay there ... At times, it comes and it can take about eight hours before it drains away.

The immediate fear was death due to drowning in the floodwaters. However, many secondary effects were dangerous for residents as well.

Every participant in the study spoke about their fears of death from disease due to the floods. Diseases attacked from multiple angles and all had the risk of death due to the diseases themselves as well as from secondary infections. Skin rashes were typical after exposure to tainted water. Malaria rates increased after stagnant waters became breeding grounds for mosquitoes. Lung diseases were frequent, likely from the microorganisms that thrive in dampness.

Gastrointestinal problems were the most universal sickness due to an exceptionally dreadful impact of the floods. To contain human waste, Jam City used pit toilets. The floodwaters overflowed the pits, flushing the waste out of them. The floodwaters then carried the human waste, spreading it throughout the community and into residents' houses. The participants explained:

> All that drains out and gets into the houses ... After that, we usually have several diseases. You find that after the floods, almost three quarters of the people are sick ... Some have passed away because of that dirty water .... Some cannot afford disinfectants. And others, you will find there was food in the house ... They have nothing else so you will have to cook the food still.

This quote also exemplifies an underlying force within the community: their poverty. The diseases could be treated. The waste and water could be cleaned. However, most of the participants did not have the means to do so.

Participants explained that they attempted to support everyone who was affected in all ways that they could because as individuals, they did not have the resources to survive. For example, the chief explained, 'When there is a sick person, we collect some money and get them to the hospital.' However, this caused deepening community-wide poverty as savings dwindled after each flood and the ensuing emergencies. When floods occurred in rapid succession, as they often did during the rainy season, there was not enough money to support all of the residents in need after each flood.

Participants also became poorer individually after the floods due to loss of their assets. One resident explained:

> You find there is no time to take anything from the house. Just be safe. You just have to walk out and leave everything in the house. So after the floods were over, usually you find nothing in the house that you can save ... So we usually, we are left, totally, without anything.

Another participant added, 'To start again a life is very difficult with nothing.'

Underlying Forces of Power

Jam City was one of the few communities in the wider study that attempted to pressure their local authorities for favourable outcomes for the community. Before the impacts of floodwall were evident, Jam City residents hoped for a voice in the development project. They did not seek to stop the construction of the apartments but sought the possibility of employment at the construction site. They reported that there was a time when government officials visited the area and construction of the apartments paused for negotiations. However shortly thereafter, construction resumed and government officials stopped returning their phone calls, which indicated a payoff.

The lack of consideration for good employment at the site was painful for the community. Instead, they were offered the worst and the lowest paid jobs, which were unacceptable to the participants. They explained:

> The people they employ from the areas are very few, very few. And only for the minor jobs like sweeping, clearing some bushes. The worst jobs they give to the people from these areas .... And if they employ us in the constructions, they pay very little amount ... They are being paid at [$3 USD] per day. From 7:30 to 5pm! [$3 USD]! No lunch. No break.

Following these interactions, they were acutely aware of their low position within society. They explained:

> And you know, they think we are thieves. They cannot live with people such as us. They usually check with the local people. They even say because of the population of the area, we will steal in the area so they have to build a wall to protect themselves. You see, they think we are just a dirty village.

The lack of power of Jam City residents was also evident in the formal responses to life-threatening floods. They received no assistance from the government, which had disaster response capabilities and residents believed should assist them in emergencies. A Jam City resident told of the lack of action by the government:

> If there's the flood, we usually call them, the governors, counsellors, they live nearby. They receive the phone calls and they don't come. They just say, we're coming, we're coming, but they usually come after the floods are over ... They will not risk themselves.

The only response local authorities ever gave Jam City community members was a demand that they relocate to another area.

The Kenyan Red Cross (KRC) brought small amounts of aid, like blankets and food, on occasion in response to the floods. However, this aid was not enough for all of the survivors. Thus, the KRC picked who would receive aid (children and the elderly), without input from the community members. The chief's wife explained:

> At times, the Red Cross people come to the areas. They bring some food, beddings, but they cannot cover for the whole village. They just give several, not to everybody. Because you find that almost everybody is affected so they just look for the elderly, the young people, they just help some, not all.

Red Cross aid is typically given in this manner. However, study participants viewed the Red Cross as a wealthy Western organization that should have enough aid for everyone. Thus, participants spoke of this angrily and declared that, 'the Red Cross does not care about us.'

Due to Jam City residents' poverty and lack of available resources, lack of action by the government and lack of charitable support, none of the participants could describe actions that promoted recovery. When asked explicitly how they recovered, one resident explained, 'We have not recovered even yet! You only survive. That's it.'

Discussion

The aim of this study was to explore the applicability of the environmental justice framework in an international context. Drawing from the literature, the principles of the environmental justice framework explored included disproportionate environmental burdens in marginalized communities, lack of participation from these communities in decisions that affect their environments, and unequal access to the benefits from the environmental hazard (Minkler et al., 2008; Nixon, 2011, Robbins, 2014; Sikor & Newell, 2014). These principles were supported in the Jam City, Nairobi, Kenya context.

Jam City residents explained they lived in the community for generations in a harmonious relationship with the environment as they purposefully lived outside of the floodplain. The new luxury buildings were built without community input, even after the community asked for their inclusion through employment at the construction site. This, in itself, is an environmental injustice due to lack of inclusion in the decision-making processes around change that affected their community and lack of access to benefits (employment) from the change. However, by building the floodwall, which channelled floodwaters directly into the neighbourhood, environmental injustice was compounded. The impacts of the floods were catastrophic and lead to sickness, death and increased poverty.

Implications for Environmental Justice

The Jam City case also adds nuances to the environmental justice framework. Participants reported that they survived the floods due to support from each other. They rescued each other from the floodwaters as well as contributed financially to the flood-induced medical crises. These actions were supported by the literature on cooperation and disasters, which theorizes that poor communities must use each other to survive disasters (Dominelli, 2012). However, scholarship has also found that many poor communities in developing countries have few resources and thus have low capacities for response to environmental degradation and are already facing countless other changes (World Bank, 2012). Thus, communities like Jam City are not able to recover through their own means. Further, the lack of formal support was illustrated in this case. This combination of lack of ability to recover from the environmental degradation and lack of formal assistance exacerbates the environmental injustice of the built apartments and resulting floods. Thus, these findings encourage the inclusion of assessments of formal supports within the environmental justice framework to explore whether injustice is further perpetuated by lack of support.

In addition, scholarship on environmental justice largely focuses on the intentional decisions to house noticeable environmental burdens in disadvantaged communities and the effects of such decisions. However, this assumes that all environmental hazards are readily identifiable as potentially dangerous, like toxic waste sites. Further, scholarship on environmental justice focuses on exploitative intent. Environmental hazards are located in marginalized communities because the people in the communities are not valued. These assumptions are not inclusive of the Jam City context. The development of new apartments is not an inherently dangerous environmental hazard. The building of a floodwall to protect the apartments was not constructed out of malicious intent. However, these actions had real and deadly consequences for the residents of Jam City. To be inclusive of the Jam City case and other similar cases, the framework of environmental justice must include non-inherently dangerous and non-intentional environmental changes.

Implications for Social Work Education

Environmental social work scholars have argued for the development of practice techniques and curriculum changes to deal with issues raised by environmental degradation in local communities (Coates, 2003; Dominelli, 2012; IFSW, IASSW, & ICSW, 2012). However, students report a lack of inclusion of these topics within their curriculum, despite their passion for the subject (Miller & Hayward, 2014).

All social workers can bring environmental degradation into their work as it impacts both micro and macro level social work and the case of Jam City can be used as an educational tool. Incorporating environmental scholarship, like this case, into social work curriculum will increase environmental understanding (Coates, 2003). The curriculum can then progress to include the physical environment in community assessments and within the eco-maps of client systems (Miller & Hayward, 2014). These assessments include environmental injustices that are not overt, like those in this case, but can be discovered by speaking with community members.

However, social work is not well established within Kenya, thus limiting the efficacy of these actions within the Jam City context. Indeed, there were no social workers in Jam City and many participants could not describe the social work profession. Thus, social work education must promote creative solutions for the Jam City neighbourhood and similar communities.

Community leadership is a skill that can be developed in socially excluded communities like Jam City that are affected by environmental injustice with little resources. The future leaders of environmental justice in Kenya do not need to be formally educated social workers. They can be local community members from marginalized areas that are interested in these issues.

Although the number is limited, there are social workers in Kenya (Mungai et al., 2014). There are likely also international social workers and development workers in the country due to the immense aid presence (Global Humanitarian Assistance, n.d). Thus social work education should include understanding the intersecting forces of environmental degradation and social justice in international contexts through exploring cases such as the case offered in this study. They can then combine their knowledge on environmental justice with their macro-centred education to support affected communities.

All of these social workers, both Kenyan and international, can support an environmental justice movement in developing countries by training interested community members in topics like community organization and community development. Communities can then use their strength, each other, in an organized and effective manner. With their people power, local communities can pressure local authorities, NGOs, their national government and the global power structure to respond to their needs and remedy environmental injustice.

Conclusion

Assessing the applicability of the frame of environmental justice in developing countries is important to build a worldwide movement for environmental fairness and structural equality. This study found that the principles of environmental justice hold in Jam City, a poor community in Nairobi, Kenya. In addition, this study adds to environmental justice scholarship through several findings. The lack of resources in Jam City limited participants' ability to respond and formal assistance was not sufficient, further perpetuating the injustice. Thus, assessments of environmental justice should include available responses to holistically understand the context.

The Jam City case also highlights a need to nuance the identification of hazardous scenarios as the causes of the environmental injustice in the community were not inherently dangerous and were not developed due to malicious intent but lead to unforeseen but treacherous consequences in the community. Environmental injustice scenarios must also include assessments of these hidden injustices. This case can be used as an educational tool to nuance environmental justice as well as to explore the principles in an international setting. Such expanded curricula can help enhance the understanding of environmental justice and further build the movement.

Disclosure statement

No potential conflict of interest was reported by the author.

References

Agyeman, J. (2014). Global environmental justice or Le droit au monde? *Geoforum*, *54*, 236–238. doi:10.1016/j.geoforum.2012.12.021

Baker, S., & Edwards, R. (2012). How many qualitative interviews is enough? Expert voices and early career reflections on sampling and cases in qualitative research. *National Centre for Research Methods*, 2–42. Retrieved from http://eprints.ncrm.ac.uk/2273/4/how_many_interviews.pdf

Coates, J. (2003). *Ecology and social work: Toward a new paradigm*. Halifax: Fernwood Press.

Coates, J., & McKay, M. (1995). Toward a new pedagogy for social transformation. *Journal of Progressive Human Services*, *6*, 27–43. doi:10.1300/J059v06n01_03

Corbin, J., & Strauss, A. (2008). *Basics of qualitative research*, 3rd ed. Thousand Oaks, CA: Sage.

Council on Social Work Education (CSWE). (2014). *Draft 3 of the 2015 Educational Policy and Accreditation Standards (EPAS)*. Retrieved from http://www.cswe.org/File.aspx?id=76675

Dominelli, L. (2012). *Green social work: From environmental crises to environmental justice*. Malden, MA: Polity Press.

Global Humanitarian Assistance. (n.d.). Kenya. *Global Humanitarian Assistance*. Retrieved from http://www.globalhumanitarianassistance.org/countryprofile/kenya

Hornsby, C. (2013). *Kenya: A history since independence*. New York: I.B. Tauris.

IFSW, IASSW, & ICSW. (2012). *The global agenda for social work and social development: Commitment to action*. Retrieved from http://www.cswe.org/File.aspx?id=60880

Kinney, P. L., Gichuru, M. G., Volavka-Close, N., Ngo, N., Ndiba, P. K., Law, A., ... Sclar, E. (2011). Traffic impacts on PM2.5 air quality in Nairobi, Kenya. *Environmental Science & Policy*, *14*, 369–378. doi:10.1016/j.envsci.2011.02.005

Kütting, G. (2004). Environmental justice. *Global Environmental Politics*, *4*, 115–121. doi:10.1162/152638004773730248

Maathai, W. (2009). *The challenge for Africa*. New York: First Anchor Books Edition.

Makworo, M., & Mireri, C. (2011). Public open spaces in Nairobi City, Kenya, under threat. *Journal of Environmental Planning and Management*, *54*, 1107–1123. doi:10.1080/09640568.2010.549631

Martinez-Alier, J. (2014). The environmentalism of the poor. *Geoforum*, *54*, 239–241. doi:10.1016/j.geoforum.2013.04.019

Miller, S. E., & Hayward, R. A. (2014). Social work education's role in addressing people and a planet at risk. *Social Work Education*, *33*, 280–295. doi:10.1080/02615479.2013.805192

Minkler, M., Vasquez, V. B., Tajik, M., & Petersen, D. (2008). Promoting environmental justice through community-based participatory research: The role of community and partnership capacity. *Health Education & Behavior*, *35*, 119–137. doi:10.1177/1090198106287692

Mohai, P., Pellow, D., & Roberts, J. T. (2009). Environmental justice. *Annual Review of Environment And Resources*, *34*, 405–430. doi:10.1146/annurev-environ-082508-094348

Mungai, N. W., Wairire, G. G., & Rush, E. (2014). The challenges of maintaining social work ethics in Kenya. *Ethics and Social Welfare*, *8*, 170–186. doi:10.1080/17496535.2014.895401

National Coordinating Agency for Population and Development (NCAPD). (2010). *Rapid Population and Development*. Retrieved from http://www.healthpolicyinitiative.com/Publications/Documents/1095_1_Kenya_booklet_3_31_10_sglepg_LOW_RES_PDF_for_INTERNET_acc.pdf

Ndege, P. (2009). *Colonialism and its legacies in Kenya*. Lecture delivered during Fulbright-Hayes group project abroad program: July 5th to August 6th 2009 at the Moi University Main Campus. Retrieved from http://international.iupui.edu/kenya/resources/Colonialism-and-Its-Legacies.pdf

Nixon, R. (2011). *Slow violence and the environmentalism of the poor*. Harvard: Harvard University Press.

Oxfam GB Kenya Programme. (2009). Urban poverty and vulnerability in Kenya: Background analysis for preparation of an Oxfam GB urban programme focused on Nairobi. *Oxfam*, 1–61. Retrieved from http://www.irinnews.org/pdf/urban_poverty_and_vulnerability_in_kenya.pdf

Padgett, D. (2008). *Qualitative methods in social work research* (2nd ed.). Thousand Oaks, CA: Sage.

Peeters, J. (20122012). The place of social work in sustainable development: Towards ecosocial practice. *International Journal of Social Welfare*, *21*, 287–298. doi:10.1111/j.1468-2397.2011.00856.x

Population Action International (PAI), *Population, climate change, and sustainable development in Kenya*. Retrieved from http://pai.org/wp-content/uploads/2013/01/Kenya_PIB_Final.pdf

Ravi Rajan, S. (2014). Environmental justice in India. *Environmental Justice*, *7*, 115–116. doi:10.1089/env.2014.7502

Robbins, P. (2014). Cries along the chain of accumulation. *Geoforum*, *54*, 233–235. doi:10.1016/j.geoforum.2012.12.007

Rogge, M. (1994). Environmental injustice: Social welfare and toxic waste. In M. Hoff & J. McNutt (Eds.), *The global environmental crisis: Implications for social welfare and social work* (pp. 53–74). Burlington, VT: Ashgate.

Saro-Wiwa, K. (1995). *A month and a day*. New York: Penguin Books.

Sepulveda Carmona, M. (2013). *Report on the special rapporteur on extreme poverty and human rights, Magdalena Sepulveda Carmona*. United Nations: Human Rights Council.

Shiva, V. (2005). *Earth democracy: Justice, sustainability, and peace*. Brooklyn: South End Press.

Sikor, T., & Newell, P. (2014). Globalizing environmental justice? *Geoforum*, *54*, 151–157. doi:10.1016/j.geoforum.2014.04.009

Silver, P. (1994). Occupational disease and the disadvantaged worker: A mandate for social work. In M. Hoff & J. McNutt (Eds.), *The global environmental crisis: Implications for social welfare and social work* (pp. 170–198). Burlington, VT: Ashgate.

Soine, L. (1987). Expanding environment in social work: The case for including environmental hazards content. *Social Work Education*, *23*, 40–46.

Stretesky, P. B., Huss, S., Lynch, M. J., Zahran, S., & Childs, B. (2011). The founding of environmental justice organizations across U.S. Counties during the 1990s and 2000s: Civil rights and environmental cross-movement effects. *Social Problems*, *58*, 330–360. doi:10.1525/sp.2011.58.3.330

Strong, D., & Hobbs, K. (2002). Administrative responses to environmental racism. *International Journal of Public Administration*, *25*, 391–417.

United Nations Development Program (UNDP). (n.d.). About Kenya. *UNDP in Kenya*. Retrieved from http://www.ke.undp.org/content/kenya/en/home/countryinfo/

Wa Kinyatti, M. (2010). *History of resistance in Kenya: 1884–2002*. Lexington: Maina wa Kinyatti.

WaterAid. (n.d.). *Kenya*. Retrieved from http://www.wateraid.org/us/where-we-work/page/kenya

Watsa, K., & Warutere, P. (2013, July 23). World Bank to help Kenya build a national safety net to reach 3.3 million poor people by 2017. Retrieved from http://www.worldbank.org/en/news/press-release/2013/07/23/world-bank-help-kenya-build-national-safety-net

Willett, J. (2015). The slow violence of climate change in poor rural Kenya communities: 'Water is life. Water is everything'. *Contemporary Rural Social Work*, *7*, 39–55.

World Bank. (2012). Turn down the heat: Why a 4° warmer world must be avoided. *A Report for the World Bank by the Potsdam Institute for Climate Impact Research and Climate Analytics.* Retrieved from http://climatechange.worldbank.org/sites/default/files/Turn_Down_the_heat_Why_a_4_degree_centrigrade_warmer_world_must_be_avoided.pdf

Wrong, M. (2009). *It's our turn to eat: The story of a Kenyan whistle-blower.* Britain: Fourth Estate.

Transforming Social Work Curricula: Institutional Supports for Promoting Sustainability

Kelly Melekis & Valerie Woodhouse

Due to the centrality of human rights and social justice in the profession, social work must collectively take a stand on ecological degradation and the climate crisis. Overall, social work education's engagement with issues of sustainability and an expanded ecological justice perspective has been peripheral and piecemeal. It is crucial that social work education expand opportunities to prepare social workers to respond to the global environmental crisis. This conceptual paper considers strategies for integrating environmental content in social work curricula and addresses the essential role of institutional supports for curricular change and professional development focused on sustainability. The role of institutional supports to advance curricular change to respond to environmental crises and promote sustainability is highlighted, along with specific examples from the authors' home institution, including (1) institutional commitment and culture, (2) curricular supports and guidelines, and (3) interdisciplinary faculty development.

> No matter what its historical role has been, social work does not have to support the existing paradigm. Because of its uniquely preserved wisdom about the radical potential of human beings to recreate themselves, social work can take a leading role in articulating a new view. (Weick, 1987, p. 228)

Since the emergence of social work, theoretical frameworks and approaches have evolved and adapted to respond to inequality and discrimination, the ever-changing political and global climate, budget cuts, changes in services, and the dynamic needs of clients and communities. However, for the past century, scientists and researchers have nervously recorded even larger changes, irrevocable and global changes, that will impact the planet and humanity for generations to come: changes in gas levels, global temperatures and sea levels rising, heat waves and cold snaps, floods and droughts, the extinction of species and shortages of food, air and water pollution, and weather patterns that are simultaneously less predictable and more severe (American Institute of Physics, 2010). Although this is a universal crisis, the devastation, trauma, and negative impacts disproportionately affect the marginalized and oppressed populations that social workers serve—people in poverty, people of color, women, indigenous people, low-income urban neighborhoods, rural communities, and older populations (Hetherington & Boddy, 2013). As environmentalist and author, McKibben (2010), implores:

> We need now to understand the world we've created, and consider urgently how to live in it. We can't simply keep stacking boulders against the change that's coming on every front; we'll need to figure out what parts of our lives and our ideologies we must abandon so that we can protect the core of our societies and civilizations. There's nothing airy or speculative about this conversation; it's got to be uncomfortable, staccato, direct. (p. xiv)

Due largely to its 'uniquely preserved wisdom' (Weick, 1987, p. 228), social work is well situated to join this conversation and support a paradigm shift toward a just and sustainable world. However, social work has made limited contributions to the global conversation about climate change or environmental degradation. Coates (2011) contends 'the large majority of social workers, like our society more generally, take "nature" for granted and have not been sensitive to the relevance of environmental issues,' particularly for social welfare, human rights and social justice (p. 1). A nationwide survey of over 400 members of the National Association of Social Workers in the US found that social workers are no more, or less, environmentally knowledgeable or friendly than the general population and the profession has not embraced the need to incorporate environmental concerns into social work education or practice (Shaw, 2013).

While increasing attention has been paid to the relevance of environmental concerns for social work, a number of scholars have suggested a paradigm shift that views 'environment' in social work more holistically rather than socially (Besthorn & Canda, 2002; Coates, 2003) and others have suggested a transnational overhaul of social work education (Hayward, Miller, & Shaw, 2012). Due to the centrality of social justice in the profession and in light of the earth's ecological crisis, it is crucial that social work education expand opportunities to prepare social workers to respond to the global environmental crisis. This conceptual paper considers strategies for integrating environmental content in social work curricula and addresses the essential role of institutional supports for curricular change and professional development focused on sustainability.

The Language of 'Green' Social Work

Any conversation about social work and the environment must take into account variations in language and the impact of language and terminology on the discourse of 'green' social work. Language around climate change and the growing ecological crisis is complex and a wide range of terms is used to describe concern for the environment, the impacts of environmental crises, and response strategies. Indeed, many concepts embedded in this discussion, such as 'sustainability' and 'social justice,' are multifaceted, complex, and socially constructed—resulting in challenges to defining and distinguishing among terms. While a history of environmental and environmental justice movements, the development of a sustainability paradigm, and explication of the nuances of various terminology is beyond the scope of this paper, we will provide some distinctions for context and consideration.

Early environmentalism focused on wilderness protection and wildlife preservation. These remain the goals of many environmental organizations today (Sandler & Pezzullo, 2007), however 'environmentalism' has received criticism for adapting into a more anthropocentric platform that does little to make real change in the climate crisis. There is concern that environmentalism 'facilitates domination of nature by developing techniques for diminishing the hazards caused by domination' (Bookchin, 1980, as cited in Ungar, 2002, p. 485). While many equate environmentalism and environmental justice, the two movements have a history of conflict and division. Proponents of environmental justice have criticized the mainstream environmental movement for engaging in racism and elitism, and valuing wilderness over people (Sandler & Pezzullo, 2007). In the USA, the Environmental Protection Agency (EPA) (2015) has defined environmental justice as:

> the fair treatment and meaningful involvement of all people regardless of race, color, national origin, or income with respect to the development, implementation, and enforcement of environmental laws, regulations, and policies. EPA has this goal for all communities and persons across this Nation. It will be achieved when everyone enjoys the same degree of protection from environmental and health hazards and equal access to the decision-making process to have a healthy environment in which to live, learn, and work. (www.epa.gov/environmentaljustice)

This definition has been criticized for failing to recognize the significance of caring for the environment while addressing structural inequalities (Dominelli, 2012). As conceptualized by Bullard and Johnson (2000), an environmental justice framework attempts to 'uncover underlying assumptions that contribute to and produce differential exposure and unequal protection' and 'rests on developing tools and strategies to eliminate unfair, unjust and inequitable conditions and decisions' (p. 559). Similarly, Schlosberg (2007) represents environmental activism in his contention that environmental justice includes the tenets of (1) fair and equitable distribution of environmental risks and benefits, (2) full participation of local residents in decisions about their environment, (3) acknowledgement of local community cultural traditions, knowledge, and ways of life, and (4) recognition of the ability of individuals and communities to make their own choices and succeed within their societies.

Some use the terms environmental justice and ecological or ecological justice synonymously, however for many proponents of ecological justice, 'the environmental justice movement is profoundly anthropocentric and is, at best, a *shallow* form of justice' (Besthorn, 2013, p. 39). Many ecologists and ecological justice activists link the climate crisis with the global obsession with consumption and economic growth, and seek a deeper and ecologically informed conceptualization of justice. Sachs (1999) suggests that ecology has turned into a worldview and that ecological thinking is relational and connective; a holistic and organicist way of viewing the world in contrast to the dominant mechanistic view that dominates modern thought (Sterling, 2009). In particular, deep ecology has been proposed as 'both a paradigm for understanding the world and a social movement for environmental responsibility' (Besthorn, 2002, p. 84). Rooted in belief that the personal and the planetary are connected, it incorporates a balance with nature, focusing on understanding our place and interdependency with nature as well as with others. As Ungar (2002) explains, 'there is a comfortable fit between the science of ecology and a profession like social work, which has as its expressed purpose fostering healthy and interdependent transactions between persons and their environments' (p. 481). As practitioners seek to address the current crisis and expand inquiry and dialog regarding responsible actions and institutions, deep ecology could facilitate a fundamental revisioning of the human/nature relationship in social work (Besthorn, 2002).

An increasingly popular term is sustainability, which has been described as a major paradigm shift and even as the 'Sustainability Revolution' (Edwards, 2005). While there is no universal definition of sustainability, many current conceptualizations are rooted in the United Nations Brundtland Report (1987) definition of sustainable development as 'a form of development that meets the needs of the present without compromising the ability of future generations to meet their own needs.' Most definitions incorporate a focus on the intrinsic interaction among ecology/environment, economy/employment, and equity/equality, often referred to as the 'Three Es' (Edwards, 2005). For the purposes of this paper, we primarily use the term 'sustainability' as we perceive it to be inclusive, multidisciplinary, and reflective of the intrinsic connection of environmental, economic, and social well-being. When using the term, we refer to the conceptualization proposed by Dillard, Dujon, and King (2009):

> Sustainability is thought of as comprised of three overlapping, mutually dependent goals: (a) to live in a way that is environmentally sustainable or viable over the long term; (b) to live in a way that is economically sustainable, maintaining living standards over the long-term, and (c) to live in a way that is socially sustainable, now and in the future. (p. 2)

We acknowledge that 'words create worlds' (Hartman, 1991), hold the tension that the language of sustainability carries, and promote transparency and critical thinking around the use of various terms as social work engages around this essential issue.

'Greening' Social Work Practice and Education

Addressing environmental concerns and incorporating sustainability into social work practice and education is not only appropriate, but imperative given our professional mission. Perhaps a tall order in our inequitable world, social work is dedicated to upholding human rights and social justice. However, social work tends to emphasize endogenous vulnerabilities and this focus on individual characteristics can 'heighten the likelihood of people being blamed for their vulnerabilities and being held accountable for a failure to adapt' (Dominelli, 2012, p. 68). Although social workers are proactively engaged in working with the most disadvantaged and marginalized members of our society, 'they need to educate themselves and others about climate change and the opportunities for micro and macro practice interventions' (Hetherington & Boddy, 2013, p. 54). There are immediate needs that social workers should be addressing, such as natural disasters, food insecurity, affordable and weatherized housing, environmental refugees, and public health crises. Poor and marginalized populations are disproportionately impacted by environmental injustices and natural disasters (Escobar, 1998) and generally face challenges to mitigating risks and dealing with the aftermath of environmental crises (Dominelli, 2012). Tester (2013) highlights the result after hurricane Katrina, where the hardest hit New Orleans neighborhoods were low-income, African-American neighborhoods located in flood zones. When such disasters occur, low-income communities have the least resources to evacuate and respond to devastation (Hetherington & Boddy, 2013). Social work has a long history of responding to social and economic challenges with an appreciation for environmental context. The need to expand our understanding of environmental issues and injustices, and enhance our practices to address them, is of paramount concern to social work education and practice today.

Expanding Person-in-Environment Perspective

While social work has long engaged with the 'environment' and has defined itself largely by its understanding and application of systems theory, the profession has centered attention on sociocultural or psychosocial environment, with an almost complete disregard of the natural environment (Besthorn & Saleebey, 2003; McKinnon, 2008). To start addressing environmental concerns and sustainability in social work practice and education, 'we should first look to broaden our view of person-in-environment to include a more holistic view of human behavior and the inclusion of the natural world' (Coates, 2003, p. 178). The person-in-environment perspective gives social work great potential to address environmental issues, however it 'limits research to person-centered, anthropogenic, and egocentric explorations' (Taylor, 2013, p. 88). Zapf (2010) refers to this perspective as 'social work's Trojan horse,' explaining that 'the words and grammatical structure suggest a dominant/subordinate relationship between two separate and distinct entities' (p. 38). Many argue that the person-in-environment perspective could benefit from an ecocentric focus, acknowledging the interdependence between humans and nature (Coates, 2003; Hoff & Polack, 1993; Mary, 2008; Zapf, 2010). Hoff and Polack (1993) encourage an ecological focus, arguing:

> The ecological model of science emphasizes that human beings are organic, intrinsically embedded in the larger ecosystem. We are in relationship and exchange with the rest of nature. Life is evolutionary, dynamic, and adaptive. From this wholistic (sic) perspective we affect and are affected by our environment. (p. 207)

Capra (2005) states that 'understanding the principles of ecology requires a new way of seeing the world and of thinking—in terms of relationships, connectedness, and context—that goes against the grain of traditional Western science and education'

(p. 20). Fortunately, social work education and practice does not require a major perceptual shift to incorporate principles of ecology or sustainability, since an appreciation for the nonlinear and relational nature of living systems is embedded within most curricula. Besthorn (2003) offers a number of pedagogical exercises for assisting students in development of ecological awareness, understanding the relationship between ecology and social justice, and articulating the role of power and politics.

Environmental Social Work Opportunities

Social workers are increasingly called upon to engage in environmental social work and may be involved in an array of issues in response to climate change and the ecological crisis. Traditional social work roles include addressing resource depletion, air and water pollution, and species extinction, supporting communities affected by natural disasters, establishing 'green' policies, providing crisis intervention and engagement skills, and emergency health services (Hetherington & Boddy, 2013). In addition, social workers engage in education, intervention, and prevention related to toxic exposure, serve on committees for policy change, monitor enforcement of environmental protection, and provide nature-based counseling and therapy (Coates, 2003; Gray, Coates, & Hetherington, 2013). Increasingly, social workers are taking a less traditional role in their fight for a livable planet. Community organizing, guerilla gardening, and social activism are examples of purposeful community change work (Shephard, 2013). These approaches utilize locality development, social planning, and social action to address issues within a community as well as the broader and systemic issues of climate change (Shephard, 2013).

Shifts in roles and responsibilities related to environmental social work and sustainability present significant implications for social work education. Very few social work schools or programs offer a focus in environmental social work, and only a small number offer content on environment, sustainability, and/or eco-social issues. Overall, social work education's engagement with issues of sustainability and an expanded ecological justice perspective has been peripheral and piecemeal. Education can be viewed as the catalyst for understanding the dynamic nature of the interrelationship between ecology, economy, and equity (Edwards, 2005). Given the current state of the global environmental crisis, immediate and direct actions are in order. The presence of institutional supports can serve as a significant facilitator for transforming social work curricula to promote sustainability.

Promoting Sustainability in Our Institutions and Programs

Institutional and programmatic factors play an essential role in the options and opportunities for infusing social work curricula with environmental and sustainability content. In this section, we will highlight several factors influencing the development and integration of such content at the authors' home institution, a public university in the northeastern USA, including (1) institutional commitment and culture, (2) curricular supports and guidelines, and (3) interdisciplinary faculty development.

Institutional Commitment and Culture

A central component of environmental or 'green' social work is a 'commitment to integrating considerations of the physical environment, the biosphere and human needs' (Dominelli, 2012, p. 126). Institutional commitment is fundamental to the advancement of sustainability within higher education. At the authors' institution, a critical element of the University's vision is its commitment to the environment and the goal of educating all of its students to understand and contribute to the sustainability of human society. As a result, the University has invested significant resources in establishing an infrastructure for environmentally related research, education, service and experiential learning. At the University level, an interdisciplinary *Envisioning Environment Work Group* was formed in 2012 to develop an inventory and recommendations regarding environmental research, education, and outreach. At the Departmental level, Social Work formed a *Sustainability Subcommittee* to ensure engagement in activities to promote sustainability.

While social work schools and programs articulate a general commitment to social justice and human rights, this is not always reflected in specific ways throughout curriculum and does not generally translate to attention to issues of environmental justice. Underlying the mission and goals of the authors' Department of Social Work is a set of core, interrelated beliefs that orients the way in which we understand, teach, practice and write about social work and sustainability: human rights, social justice, a strengths perspective that emphasizes resilience and rebound, and critical social construction. In response to the exigencies of our times, our social work faculty undertook intensive curriculum redevelopment to produce theoretically rich, contextually responsive curricula. As a result, the department has culled a niche in an area of concentration we call *Transformative Social Work*. The skills in deconstruction and relational, generative, and profound change integral to this curriculum are directed at oppressive discourses and injurious relationships and practices of power from their most intimate to their most public. The urgency our faculty feel about this work is due to recognition that these discourses, relationships and practices are unsustainable for the earth and we her people. The transformative social work concentration is in part a response to the belief that we have reached a point where incremental change is not enough. This existing infrastructure provides a venue for a curricular response to the current emergency in our natural and social environment. Where infrastructure does not exist, it would be important to develop organizational structures to support a commitment to environment within social work education.

Curricular Supports and Guidelines

In a recent study of US social work students' attitudes, interests in, and practices related to environment, Miller and Hayward (2014) found that a majority view environmental justice as an important aspect of social justice and area of concern for social workers, and expressed interest in more environmental justice content in social work education. One of the ways in which our institution is integrating its sustainability vision across curricular and co-curricular activities is through its General Education Initiative. The University's Student Government Association (SGA) introduced and passed a resolution supporting the creation of a general education requirement concerning sustainability for all undergraduates. The Faculty Senate then approved a resolution to (1) adopt Sustainability Learning Outcomes (SLOs) (see Table 1) and (2) create a Sustainability Curriculum Review Committee. As a result, whatever their chosen discipline, each undergraduate student will be expected demonstrate their mastery of the defined learning outcomes in the *knowledge, skills* and *values* categories, as well as the *personal domain*.

The lead author participated in a curriculum-mapping workshop to identify where SLOs are currently addressed in the undergraduate social work core curriculum, potential core courses for infusing SLO, and possible mechanisms for assessing SLOs. Based on a review of core course syllabi for the social work major and inquiry to all core course instructors, the lead author identified courses where SLOs are current included, as well as those courses where SLOs could potentially be infused. While this work represents only the beginning of the process (Uchiyama & Radin, 2009), and is specific to the authors' institution, it is evident that curriculum mapping could be a beneficial start to transforming social work curricula to address environmental and sustainability content. In addition, this ongoing, dynamic process has the potential to increase engagement and foster collaboration among colleagues.

There are a number of potential strategies to support curricular change and promote sustainability within social work education. Despite varied opinions about accreditation and degrees of controversy surrounding curriculum policy and accreditation standards (Robbins, 2014), it may be worthwhile to consider how sustainability learning objectives complement existing guidelines and whether or where they might be included in educational and professional standards, such as the Council on Social Work Education's (CSWE) Educational Policy Accreditation Standards (EPAS) in the US The Association for the Advancement of Sustainability in Higher Education (AASHE) curriculum working group (2009) developed a framework to assist in integrating sustainability across the curriculum in higher education, including general principles for sustainability education, specific educational goals, and representative student learning outcomes for assessment within a course or assignment. Besthorn (2003) and Besthorn and Canda (2002) have

Table 1 Sustainability Learning Outcomes (SLOs)

	Learning outcome	Domain
1	Students can have an informed conversation about the multiple dimensions and complexity of sustainability.	Knowledge
2	Students can evaluate sustainability using an evidence-based disciplinary approach and integrate economic, ecological, and social perspectives.	Skills
3	Students think critically about sustainability across a diversity of cultural values and across multiple scales of relevance from local to global.	Values
4	Students, as members of society, can recognize and assess how sustainability impacts their lives and how their actions impact sustainability.	Personal

offered specific strategies for how insights from deep ecology can be incorporated into specific social work courses and assignments. In addition to specific course adjustments, it would behoove social work programs to consider applying a broader systems perspective to engage in curriculum mapping to ensure infusion of sustainability learning objectives. Sterling (2009) suggests that curriculum designers and teachers (across disciplines) develop 'learning situations where the potential for transformative learning experiences is made manifest' (p. 82) and offers a framework for course design and pedagogic approaches to invoke 'learning as change' in the pursuit of sustainability. Jones (2010) suggests that the challenge for social work education is 'to integrate transformative learning theory with a range of existing methods focused on reflective, dialogic, and experiential approaches, and to apply this theory and method to the development of ecological awareness and literacy among students' (p. 75) and offers a framework for an ecologically transformed curriculum for social work education (Jones, 2013). With an educational structure that encourages reflexive, experiential and action-oriented learning, social work is well positioned to facilitate ecological transformation that links the personal with the environmental, and manifests in individual and political action.

Interdisciplinary Faculty Development

Inclusion of environmental and sustainability content in social work curriculum requires knowledgeable and prepared faculty. In 2009, the authors' University started a Sustainability Faculty Fellows (SFF) Program to provide faculty development opportunities in environment and sustainability. Rooted in the Faculty Learning Community (FLCs) model (Cox, 2004), the program engages small, multidisciplinary groups of faculty in a yearlong scholarly pursuit of teaching, learning and community building focused on sustainability. The focus is on core conceptual knowledge as well as pedagogical strategies. Four cohorts of faculty completed the program between 2009 and 2013, resulting in a collaborative network of faculty, inclusion of sustainability in curricula, and student engagement in sustainability.

During the 2012–2013 academic year, the lead author participated in the SFF program and infused sustainability content into a BSW Social Work Practice II course. Required of all undergraduate majors and taken concurrently with field practicum, the course focuses on organizational and community practice. A new module on Community Development and Sustainability was added to the course. Learning objectives for this module were adjusted, so that beyond learning to describe and discuss approaches to community development, students were expected to apply systems thinking to identify sustainable practices for community development and change, and discuss the role of diversity in sustainable communities. Students applied concepts from Meadows (1999) *Leverage points: Places to intervene in a system* to Hill and Marron's (2007) *Brooklyn Matters*, a documentary about the fight against the Atlantic Yards development project in Brooklyn, NY, to discuss places where intervention occurred, or could have, within various systems. Students also engaged in a case study review using Fauri and colleagues' (2008) *Riverton: Envisioning a sustainable community*, focusing particularly on the concept of sustainable community development and the linkages between sustainability and social justice. In addition, students received a guided tour of a special exhibit at the University museum, 'High Trash,' consisting of works by visual artists that use found materials to represent concerns of waste, ecology, and humans' place on this planet. Following this viewing, the class engaged in reflective discussions about the relationship between social work and sustainability, the social implications of environmental destruction and degradation, and the environmental implications of our personal and professional actions.

Faculty who are interested in environmental social work and committed to including this content in their courses would likely benefit from enhanced understanding of environmental concepts and support for professional development to explore pedagogical strategies to engage students in sustainability. Indeed, interdisciplinary faculty development facilitated the inclusion of sustainability content in our undergraduate social work curriculum. The value of interdisciplinary collaboration and networking cannot be underestimated. It provides a rich opportunity for viewing sustainability principles and considering strategies for promoting sustainability in social work education and practice.

Conclusion

The context in which social work operates has changed radically in the era of globalization and neoliberalism with significant consequences for the lives and relationships of the people social workers serve. Major shifts in world politics, national economies, and social and environmental climates compel us to critically analyse the theories and beliefs we have relied on to make sense of things and to guide our actions. How will we respond to increasing environmental degradation and the intensification of impoverishment of disadvantaged nations and groups (McNutt & Hoff, 1994)? As a profession, we are long overdue for developing a vision of what our role will be in a society plagued by environmental crises. We recognize that incremental measures to correct dramatic shortfalls and subtle alternations in policy and practices portent little help to any of us, most especially to those who suffer the indignities of intersecting oppressions. As Rogue (1993) articulates, 'the ethical commitment of the social work profession, combined with the threat to disenfranchised communities from environmental hazards, suggests a clear call to action for social work practitioners, academicians, institutions, and associations' (p. 117).

Social work education requires transformation to reflect our changing society as environmental crises have significant implications for both classroom curricula and field practice. As Edwards (2005) states, 'through education, sustainability can become firmly established within the existing value structure of societies while simultaneously helping that value structure evolve toward a more viable long-term approach to systemic global problems' (p. 23). Institutional and programmatic factors play an essential role in the options and opportunities for infusing social work curricula with environmental and sustainability content. While individual faculty committed to the cause can certainly make an impact, institutional supports strengthen the potential for an effective professional response to the current emergency in our natural and social environment. Institutional commitment and culture, curricular supports and guidelines, and interdisciplinary faculty development are factors to be considered and nurtured as we seek to prepare social workers with the knowledge, skills, and values to respond to environmental crises and promote sustainability.

Acknowledgements

The authors would like to acknowledge Emily Sanders and Mark Melekis for their assistance in reviewing drafts of this manuscript, as well as the UVM Department of Social Work Sustainability Subcommittee members: Susan Comerford, Fiona Patterson, Susan Roche, Brenda Solomon, and Jessica Strolin-Goltzman.

Disclosure statement

No potential conflict of interest was reported by the authors.

References

American Institute of Physics. (2010 May). Introduction: A hyperlinked history of climate change science. *The Discovery of Global Warming*. Retrieved from http://www.aip.org/history/climate/summary.htm

Association for the Advancement of Sustainability in Higher Education (AASHE). (2009). *Sustainability education framework*. AASHE Curriculum Working Group. Retrieved from www.aashe.org/files/documents/board/SustEd_Framework_8-5-09.doc

Besthorn, F. H. (2002). Radical environmentalism and the ecological self: Rethinking the concept of self-identity for social work practice. *Journal of Progressive Human Services*, *13*, 53–72.

Besthorn, F. H. (2003). Radical ecologisms: Insights for educating social workers in ecological activism and social justice. *Critical Social Work*, *4*, 66–106. Retrieved from http://www1.uwindsor.ca/criticalsocialwork/radical-ecologisms-insights-for-educating-social-workers-in-ecological-activism-and-social-justice

Besthorn, F. H. (2013). Radical equalitarian ecological justice: A social work call to action. In M. Gray, J. Coates, & T. Hetherington (Eds.), *Environmental social work* (pp. 31–45). New York, NY: Routledge.

Besthorn, F. H., & Canda, E. R. (2002). Revisioning environment. *Journal of Teaching in Social Work*, *22*, 79–101. doi:10.1300/J067v22n01_07

Besthorn, F. H., & Saleebey, D. (2003). Nature, genetics, and the biophilia connection: Exploring linkages with social work values and practice. *Advances in Social Work*, *4*(1), 1–18. Retrieved from https://journals.iupui.edu/index.php/advancesinsocialwork/issue/viewFile/8/8

Bookchin, M. (1980). *Toward an ecological society*. Montreal: Black Rose.

Brundtland, G. H. (1987). *Our common future: Report for the World Commission on Environment and Development, United Nations*. Retrieved from http://www.un.org/ga/search/view_doc.asp?symbol=A/RES/42/187

Bullard, R. D., & Johnson, G. S. (2000). Environmentalism and public policy: Environmental justice: Grassroots activism and its impact on public policy decision making. *Journal of Social Issues*, *56*, 555–578. doi:10.1111/0022-4537.00184

Capra, F. (2005). Speaking nature's language: Principles for sustainability. In M. K. Stone & Z. Barlow (Eds.), *Ecological literacy: Educating our children for a sustainable world* (pp. 18–29). San Francisco, CA: Sierra Club Books.

Coates, J. (2003). *Ecology and social work.* Blackpoint: Fernwood Publishing.

Coates, J. (2011). Social work and the environment: A progressive step forward. *Global Alliance for a Deep Ecological Social Work.* Retrieved from http://ecosocialwork.org

Cox, M. D. (2004). Introduction to faculty learning communities. *New Directions for Teaching and Learning, 2004,* 5–23. doi:10.1002/tl.129

Dillard, J., Dujon, V., & King, M. C. (Eds.). (2009). *Understanding the social dimension of sustainability.* New York, NY: Routledge.

Dominelli, L. (2012). *Green social work: From environmental crises to environmental justice.* Cambridge: Polity Press.

Edwards, A. R. (2005). *The sustainability revolution: Portrait of a paradigm shift.* British Columbia: New Society.

Escobar, A. (1998). Whose knowledge, whose nature? Biodiversity, conservation, and the political ecology of social movements. *Journal of Political Ecology, 5,* 53–82. Retrieved from http://jpe.library.arizona.edu/volume_5/3escobar.pdf

Fauri, D. P., Wernet, S. P., Netting, F. E. (2008). Riverton: Envisioning a sustainable community. In *Cases in macro social work practice* (3rd ed., pp. 92–106). Boston, MA: Pearson Education.

Gray, M., Coates, J., & Hetherington, T. (2013). Conclusion. In M. Gray, J. Coates, & T. Hetherington (Eds.), *Environmental social work* (pp. 46–61). New York, NY: Routledge.

Hartman, A. (1991). Words create worlds. *Social Work, 36,* 275–276. doi:10.1093/sw/36.4.275

Hayward, R. A., Miller, S. E., & Shaw, T.V. (2012). Social work education on the environment in contemporary curricula in the USA. In M. Gray, J. Coates, & T. Hetherington (Eds.), *Environmental social work.* New York: Routledge.

Hetherington, T., & Boddy, J. (2013). Ecosocial work with marginalized populations: Time for action on climate change. In M. Gray, J. Coates, & T. Hetherington (Eds.), *Environmental social work* (pp. 46–61). New York, NY: Routledge.

Hill, I., & Marron, M. (2007). *Brooklyn matters.* Harriman, NY: New Day Films.

Hoff, M., & Polack, R. (1993). Dimensions of the environmental crisis: Challenges for social work. *Social Work, 38,* 204–211. doi:10.1093/sw/38.2.204

Jones, P. (2010). Responding to the ecological crisis: Transformative pathways for social work education. *Journal of Social Work Education, 46,* 67–84. doi:10.5175/JSWE.2010.200800073

Jones, P. (2013). Transforming the curriculum: Social work education and ecological consciousness. In M. Gray, J. Coates, & T. Hetherington (Eds.), *Environmental social work* (pp. 213–230). New York, NY: Routledge.

Mary, N. (2008). *Social work in a sustainable world.* Chicago, IL: Lyceum Books.

McKibben, B. (2010). *Earth: Making a life on a tough new planet.* New York, NY: Times Books.

McKinnon, J. (2008). Exploring the nexus between social work and the environment. *Australian Social Work, 61,* 256–268. doi:10.1080/03124070802178275

McNutt, J. G., & Hoff, M. D. (1994). Dilemmas and challenges for the future of social welfare and social work. In M. D. Hoff & J. G. McNutt (Eds.), *The global environmental crisis: Implications for social welfare and social work* (pp. 297–305). Brookfield, VT: Ashgate.

Meadows, D. (1999). *Leverage points: Places to intervene in a system.* Hartland, VT: The Sustainability Institute. Retrieved from http://www.donellameadows.org/archives/leverage-points-places-to-intervene-in-a-system/

Miller, S. E., & Hayward, R. A. (2014). Social work education's role in addressing people and a planet at risk. *Social Work Education, 33,* 280–295. doi:10.1080/02615479.2013.805192

Robbins, S. P. (2014). From the editor: Accreditation, competency-based education, and EPAS revisions. *Journal of Social Work Education, 50,* 581–586. doi:10.1080/10437797.2014.947893

Rogue, M. (1993). Social work, disenfranchised communities, and the natural environment: Field education opportunities. *Journal of Social Work Education, 29*, 111–120. doi:10.1080/10437797.1993.10778803

Sachs, W. (1999). *Planet dialectics*. London: Zed Books.

Sandler, R. D., & Pezzullo, P. C. (Eds.). (2007). *Environmental justice and environmentalism: The social justice challenge to the environmental movement*. Cambridge, MA: MIT Press.

Schlosberg, D. (2007). *Environmental justice: Theories, movements and nature*. Oxford: Oxford University Press.

Shaw, T. V. (2013). Is social work a green profession? An examination of environmental beliefs. *Journal of Social Work, 13*, 3–29. doi:10.1177/1468017311407555

Shephard, B. (2013). Community gardens, creative community organizing, and environmental activism. In M. Gray, J. Coates, & T. Hetherington (Eds.), *Environmental social work* (pp. 121–134). New York, NY: Routledge.

Sterling, S. (2009). Ecological intelligence. In A. Stibbe & H. Luna (Eds.), *The handbook of sustainable literacy*. Cambridge: Green Books.

Taylor, S. A. (2013). Social science research in ocean environments: A social worker's experience. In M. Gray, J. Coates, & T. Hetherington (Eds.), *Environmental social work* (pp. 88–101). New York, NY: Routledge.

Tester, F. (2013). Climate change as a human rights issue. In M. Gray, J. Coates, & T. Hetherington (Eds.), *Environmental social work* (pp. 102–118). New York, NY: Routledge.

Uchiyama, K. P., & Radin, J. L. (2009). Curriculum mapping in higher education: A vehicle for collaboration. *Innovative Higher Education, 33*, 271–280. doi:10.1007/s10755-008-9078-8

Ungar, M. (2002). A deeper, more social ecological social work practice. *Social Service Review, 76*, 480–497. doi:10.1086/341185

U.S. Environmental Protection Agency. (2015). *What is environmental justice?* Retrieved from www.epa.gov/environmentaljustice

Weick, A. (1987). Reconceptualizing the philosophical perspective of social work. *Social Service Review, 61*, 218–230. doi:10.1086/644436. Retrieved from http://www.jstor.org/stable/30011882

Zapf, M. (2010). Social work and environment: Understanding people as place. *Critical Social Work, 11*, 30–46. Retrieved from http://www1.uwindsor.ca/criticalsocialwork/social-work-and-the-environment-understanding-people-and-place

Environmental Sustainability and Social Work: A Rural Australian Evaluation of Incorporating Eco-Social Work in Field Education

Fran Crawford, Savana Sabine Agustine[1], Leah Earle, Ahmed Bawa Kuyini-Abubakar, Yoni Luxford & Hurriyet Babacan

Climate change poses significant threat to the wellbeing of global society. Addressing this change has as yet generated no fixed blueprint for social work practice and education. This paper reports on a formative evaluation of one Australian initiative to address this transformative opening in social work field education. Prompted by service users' and workers' experience of the impact of drought, a rurally located social work course team amended the field education curriculum to include a focus on Environment and Sustainability. This learning goal was added to the existing learning goals derived from the Australian Association of Social Workers (AASW) Practice Standards. Students and field supervisors were surveyed on their experience of meeting this new learning goal. While most expressed confidence in understanding the concepts involved, they clearly lacked assurance in interpreting these in practice encounters. Considering their qualitative input suggests that this topic is making a transition from being on the margins of social work to becoming mainstream. Their open-ended responses indicate that the incorporation of environmental sustainability into practice is at a threshold stage of development. Further enactment of eco-social work at the local level is concluded to be supported by using a transformative learning framework in facilitating critical reflection and collaborative dialogue for effective change.

Introduction

Climate change is a global transformation that affects everyone and has significant implications for social work practice. The interrelationship between the environmental, social, and economic aspects of sustainability is beginning to be accepted as a starting point for action (Molyneux, 2010). This approach recognizes key social justice issues as they relate to the environment including global poverty and economic growth, environmental degradation and inequity, women's oppression and population growth, equity between generations and sustainable and inclusive development (Dominelli, 2011; Bowers, 2009). Though the science evidences a need for long-term change, political uncertainties attaching to such action have restricted commitment across many disciplines, including social work.

Rural Australians have been wrestling with the effects of greater variability in rainfall, prolonged droughts, and a greater incidence of extreme weather events for decades (Hannam, 2014, p. 5). This paper reports on a formative evaluation of a rurally-based social work education team's initiative in including an overt focus on Environment and Sustainability among the learning goals of field placement in a social work program in such a setting.

Background to Formative Evaluation

In 2009, the first cohort of social work students enrolled at a university situated on the rural tablelands of northern New South Wales (Maple, Jarrott, & Kuyini, 2013). Though the university dates from 1954, curriculum development for a social work course began in 2007 when policy-makers recognized that the lack of accessible human services compounded the rural disadvantage attaching to drought (Maple et al., 2013).

Using the university's experience in providing rurally-conscious distance education to isolated students, the new course was designed to meet a regional shortage of social workers and attract students with rural awareness (Maple et al., 2013). The course philosophy was that those trained in the bush stay in the bush.

The new course provided an opportunity 'to break with discipline-wide accepted teaching methods and models and create novel, new ways' (Maple et al., 2013, p. 350). Galvanized by the field education coordinator's attendance at a conference, Eco-Social Work: Transforming the national social work curriculum in 2013, the team took action. In July 2013, sustainability was introduced for all field placements as Learning Area 8: Environment and Sustainability: *Articulation of the interface between social work and environmental issues including an ability to uphold eco-social justice as a core obligation.*

The 2010 Code of Ethics of the AASW (the national accrediting body) highlights the physical environment as integral to wellbeing but as yet this is not specified in the AASW Practice Standards (2013) (McKinnon, 2013). Each Australian accredited course identifies field placement learning areas drawn from these Practice Standards (2013). Adding sustainability to the learning areas derived from the AASW Practice Standards resonated with growing involvement in the issue of environment within social work, such as special issues of social work journals, including the AASW journal, Australian Social Work (2013a, 2013b).

In seeking to formally evaluate how students and field supervisors experienced the addition of the eighth learning area, [Agustine and Kuyini-Abubakar] designed and administered a survey drawing on the constructivist methodological approach that informed the design of the social work program. The transformative learning and reflective practice philosophies integral to the design of the new course were used as discursive tools in the design of this study. While there were both summative and formative aspects to the survey, this paper summarizes the summative aspects and focuses on the qualitative input of students and field supervisors (Patton, 2008). The design of the survey took for granted the mutually informing relationship between practice and theory and sought to usefully improve on the university educators' first step in introducing this new learning goal (Schwandt, 2014).

Social Work Skills and Climate Change

Dominelli (2011) urges social workers to master the scientific evidence behind climate change. The science of climate change suggests that fitting individuals better into their environment no longer suffices as the dominant narrative of social work (Coates & Gray, 2012; Norton, 2012). Rather people and environment are seen to be in mutually constitutive relationship. A green social movement addresses this scientific fact. Despite a history of active relationship with social movements (Thompson, 2002), the profession of social work has been slow to officially recognize the interrelationships between the environmental, social, and economic aspects of sustainability (Coates & Gray, 2012). Jones (2011) details the case for recognizing the inextricable relationships between the concerns of ecological justice and those of social justice. Such recognition has profound implications for how the values of social work are best enacted in practice. As yet there is scant literature detailing this.

Reviewing the social work literature on eco-social work Molyneux (2010) found a lack of 'pragmatic suggestions on how to apply environmental or eco-social work in practice and few studies appeared to have explored the practical realities' (p. 61). Nineteen of the 21 articles reviewed were conceptual with only two empirical. Her work identifies a need to support the evolution of eco-social work from ideas to action.

Threshold Concepts and Liminal Space

Speaking of integrating the disabled people's social movement in social work education, Morgan (2012) recommends 'identifying what it is we want students to "get", whether they find this troubling and how they can be effectively supported as they move through liminal spaces in social work education' (p. 215). Morgan advocates developing skills (such as critical thinking and reflexivity) that allow students to envisage and enact diverse ways of achieving the purpose of social work. Taking *threshold concepts* as 'new and previously inaccessible ways of thinking about something' (Meyer & Land, 2003, p. 1), Morgan advocates using this idea to tackle emerging topics in social work education. Arguably environmental sustainability and eco-social justice are such emerging topics.

A threshold concept has five characteristics that support and scaffold a changing curriculum. It is *transformative* in involving a significant shift in insight and usually *irreversible* in that 'they are likely to change a perspective in a way that is hard to undo' (p. 219). Thirdly, threshold concepts are *integrative* in facilitating connections to be made between previously unrelated ideas and facts. The characteristic of *boundedness* refers to questions of demarcation. There may be perceived clashes or contradictions such as seeing environmental sustainability as purely scientific. Finally, threshold concepts frequently involve *troublesome* knowledge especially for students 'who take their cues from wider social and cultural depictions and scripts for understanding' (p. 220).

Concepts like environmental sustainability can 'undermine deeply-held beliefs and attitudes and ... trigger fundamental questions about the nature, practice and future existence of social work itself' (Morgan, 2012, p. 220). Educational researchers find that in enacting the threshold concept, learners occupy a liminal space (Meyer and Land, 2003; Cousin, 2006). Liminal space is a space of transition, not knowing and uncertainty. Requiring students to enact the concept removes options of remaining unaware or rejecting the idea. In response to feeling unsafe in such a situation, students may ask for clear instructions and deliver 'a ritualised performance which may only be uncovered when the performance slips' (Morgan 2012, p. 220). To forestall such ritualized performance, the transformative learning and reflective practice associated with the threshold concept become more important in how social work students and field practitioners will come to understand/conceptualize eco-social work practice in an un-demarcated landscape of practice opportunities.

The Research Design

The electronic survey, approved by the University Ethics Committee (HE 13-272), was sent to all students from the Bachelor and Masters degree completing first or second placement in 2013 ($n = 47$) and all field supervisors ($n = 39$) of the students. The same questions were asked of students and supervisors with minor rewording reflective of their differing roles. The response rate for students was 48.9% ($n = 23$) and for field supervisors 56.4% ($n = 22$).

This paper focuses on the open-ended responses reflecting beliefs, attitudes, knowledge, and attributes about environmental sustainability and social work. The closed-ended questions covered demographic factors such as age and a series of self-ratings as to the degree of knowledge held on the topic of environmental sustainability. Questions regarding environmental sustainability were:

1. On a scale of 1 to 5 (with 1 being the lowest and 5 being the highest), please rate your overall knowledge on Environment and Sustainability as a social work area of practice.

2. On a scale of 1 to 5 (with 1 being the lowest and 5 being the highest), please rate your knowledge on the following:
    (a) Role of social worker in the area of Environment and Sustainability
    (b) Eco-social Work
    (c) Integration of eco-social work in everyday social work practice
    (d) Impact of Environment and sustainability in a global context
    (e) Relationship between social justice and Environment and Sustainability
3. What did you understand on what was required of you to address learning area 8 in your field education learning plan?
4. What challenges did you face to fulfill the requirements of the learning area 8 (environment and sustainability) of your field education learning plan?
5. Suggest strategies that could be used to resolve the challenges you faced in trying to fulfill requirements of the learning area 8 (environment and sustainability) of your field education learning plan.
6. What additional training would you like to see for students in future to be able to articulate learning area 8 (environment and sustainability) in their social work practice?

Results

In this report, we highlight the results of the open-ended responses as they reflect just how students and practitioners perceived and engaged with practice in the area of environment and sustainability during field placements. A summary only is provided for the close-ended questions.

Close-ended Questions

Q1 Overall knowledge

In respondent's rating of their overall knowledge as to environment and sustainability as a social work area of practice, some 60% of both groups rated themselves at 3 or above on the five-point scale.

Q2 Sub-categories of knowledge

Responses indicate that both students and supervisors feel more confident as to their conceptual knowledge of the global connection between environmental sustainability and social work than they did about enacting that knowledge in contextualized practice.

Open-ended Questions

Analysis of students' and supervisors' open-ended responses revealed uncertainty, struggle, confusion through to excitement on how to enact Learning Area 8 in practice. Responses to each of the four open-ended questions have been clustered into themes.

Q3 What did you understand on what was required of you to address learning area 8 in your field education learning plan?

Here there were three themes identified: Limited view/ waiting to be told; Open to learning/embracing the challenge, and Taking action.

Limited view

The limited view theme of waiting to be told or defaulting to 'recycling paper' or 'not driving somewhere' is captured in student comments such as 'very little,' 'had difficulty understanding,' and 'clutching at straws'.

> Very confused with this area. My supervisor tried to help by linking with readings on green social work ... This area was my biggest struggle.

Some field supervisors also had limited understanding and struggled to support students.

- I had no idea and I think it was a perplexing question for the student too. ....it seemed like a filler topic ... not sure how useful it actually was.
- Both myself and student struggled with this area.
- No understanding of what was required.

One supervisor (a social worker with a science degree) found eco-social work a ludicrous proposition because 'social workers have no scientific knowledge.'

Open to learning

Students in this cluster 'had to do my own research' but 'by the end of the placement I did understand what was required.' A student described reflecting:

- On the importance of the natural environment as yet another vital context that is supportive or damaging to our physical/social/emotional/mental/spiritual health...

Supervisors often identified a lack of knowledge alongside an openness to learning more.

- It was a complex area to understand and was not clear until (university academic) came out and explained it.
- Led by student. We discussed topics that she raised and that she had written about. She shared her readings.

Taking action

The theme of taking action and creating awareness had fewer responses and covered 'seeing practical ways of promoting environmental awareness' and 'doing something green in order to create awareness between my clients and peers.' This last comment noteworthy as only a few responses mentioned clients and were more likely to mention agency operations:

- How the agency I was on placement with dealt with (or did not deal with) protecting and sustaining the environment.

Supervisors gave a range of responses with several focusing on the importance of supporting students to appreciate context in a culture embracing standardization in knowledge development.

- Support learning that challenges paradigms that ignore the importance of place, the role of environment in life and life's choices.
- To support the student to think about and engage in potential activities appropriate to the placement context with an eco focus.

Q4 What challenges did you face to fulfill the requirements of the learning area 8 (environment and sustainability) of your field education learning plan?

Eighteen out of twenty-three students mentioned their 'lack of understanding' of this area of practice. Students in structured settings like child protection and hospitals, struggled to identify its relevance. Four themes were repeated across both sets of responses: Lack of awareness/training; 'Stuck' on paper and recycling; Making the link to social work and Embracing the challenge.

The lack of awareness and/or training

This applied to students, supervisors, faculty, and agencies. Students wrote:

- Difficult to relate to child protection... my supervisors were not much help as they had not been trained in the ideals of it.
- My agency peers were mostly over the age of 50 and didn't appear willing to give much importance to the environment.

Supervisors' comments identified a wider range of lacks:

- Trying to understand the scope of this area and how it applies to working in an area with a welfare focus.
- Lack of prior learning on my part.
- As an agency it wasn't something we had thought about much before.
- As we are risk and crisis driven, wondered about how to engage families in environmental concerns.

Supervisors also commented on student lacks:

- Student resistance! It is a rural context so it can be more problematic.
- The biggest challenge was getting students to engage with this learning area. I think it is extremely important.

'Stuck' on recycling

This theme relates to the previously identified limited view. A student wrote:

- Many of us were looking at small scale items such as recycling in the workplace... this learning area is much bigger than that.

Supervisors' comments mirrored this:

- The student thought it was about turning the lights off and recycling.

Making the link to social work

Several students 'could not see the relevance' and 'had little knowledge of how it applied to social work practice'. For supervisors, the agency setting shaped possibilities. So one had no difficulty in making the link because 'a community development placement is easily aligned with environment and sustainability'. In contrast another had 'extreme difficulty... in a hospital setting.'

Embracing the challenge

Both students and supervisors gave positive examples of negotiating challenges.

A student reports,

> I got creative and came up with ideas, received the support of my supervisor and learnt more along the way.

Another comments that Learning Area 8 was the most enjoyable part of placement 'with the introduction of many energy saving challenges and addition of outdoor components to my agency's already strong focus on the environment.'

Supervisors similarly commented that they 'didn't find this challenging and the student made an important contribution.'

Q5 Suggest strategies that could be used to resolve the challenges you faced in trying to fulfill requirements of the learning area 8 (environment and sustainability) of your field education learning plan.

Among suggestions on resolving challenges, three themes were identified across both groups: Taking Initiative; Better Preparation; and Give Specific Guidance. Students mentioned More Course Content while some supervisors' responses generated a theme of Troublesome Territory. The responses in each category overlap with responses in other categories and all could be placed on a continuum from valuing independent and creative thinking to valuing standardized learning templates. These continuum ends reflect conceiving of knowledge as fixed and unchanging and/or constantly in flux and contested.

Troublesome territory

One supervisor wrote 'this is a wasted learning goal and it should be removed. I supervise students from many universities and no others focus on this.' Another rejected the learning area because social workers were not scientists. Another stated they could only supervise students doing casework implying that addressing environmental sustainability could not be casework. These all tended to see knowledge as fixed. Another, seeing knowledge as contested, rejected it because of community politics involved, the 'policies and ethics of the AASW' and 'a power game' between social work and those with community development qualifications.

More course content

Students generally wanted to be better prepared through course content to address this placement-learning goal. Some sought an opening up of ideas for development in practice. At the same time many students and supervisors sought a closing down of uncertainty through requesting specific guidance as to how to address the task.

Give specific examples

Some identified what they saw as the importance of standardizing what was required. One student suggested 'having a template of different ideas and guidelines that could be applied in different social work environments.'

Better preparation

Both students and supervisors wanted better preparation to engage together on meeting Learning Area 8. This included involving agency and university in supporting the change. Better communication and more discussion were also requested.

Taking initiative

One supervisor, reframing traditional teacher/learner relationships stated, 'Advise students that supervisors appreciate the opportunity to access learning materials and to hear about the topic from them.'

Students suggested self-research and if still unsure then to contact university staff. One stated,

> Do plenty of research. Think outside the square. Try and look at it as linking to all learning areas not as a separate issue.

Q6 What additional training would you like to see for students in future to be able to articulate learning area 8 (environment and sustainability) in their social work practice?

Supervisors wanted more resources to better equip them for their role. Some added that their agency did not prioritize this as core business.

> 'I am happy to do an on-line training program.'

Students focused on adding a new unit to the course.

> A unit based on eco-social work.. our environment is crucial in our wellbeing.

Others advocated scaffolding such content throughout the course.

> More information on it within other units to allow students to grasp its meaning before heading into placement unsure and confused.

Both students and supervisors named the importance of learning from peers. Exploring creative ideas was seen as helpful. While one stated 'well done, excited about the new inclusion' another stated:

> There was a lot of talking "at" students which provided valuable information but there was not the opportunity for students to bridge that learning gap and begin to test the waters and articulate their understanding thus far.

Discussion

The results strongly resonate with Molyneux's (2010) finding that eco-social work to date is more conceptual than actual. Both students and supervisors were confident in their knowledge of environmental sustainability issues in a global rather than practice context. Perhaps reflecting this lack of grounded knowledge there was little awareness of the fact that clients are often experts in living sustainably not as a life-style choice but through necessity (Gibson, Waitt, Head & Gill, 2011). Only a few responses mentioned clients and there were several responses suggesting that this learning area could only be addressed through community development approaches and not with individuals. This would suggest that in curriculum development it is important to embed the big picture from macro-practice in micro-practice. Some students and supervisors, both through the survey and more generally, were able to give feedback as to how clients valued practical support in saving on power and food bills. Yet institutional blinkers meant that other students failed to see any link between their placement in child protection and health agencies and utilizing such practical avenues to improved well-being.

Certainly there was keenness to 'test the waters' as one student expressed it. In this, there would seem to be a need to emphasize the active role of all stakeholders in dialoguing about their lived experiences of environment. Moving beyond a static conceiving of *people in environment* this would take up Zapf's (2010) reframing to *people with environment* where the aim is not to control the environment but rather to understand and interactively use the environment to better wellbeing.

Considering the spectrum of responses, the survey results suggest that the introduction of the topic of environmental sustainability in this course currently fits the five characteristics of a threshold concept (Morgan, 2012). Clearly some students found Learning Area 8 *transformative* in involving a significant shift in insight and for those doing so it was likely to involve a lasting shift.

Transformative education in relation to adult learners and social work practitioners is based on the assumption that adults go through major transformations in their life (Mezirow, 1996, 1997, 2000), and learning approaches that transform the way we think, believe, and feel about particular issues are important for innovative practice. Climate change is one universal transformation that affects all individuals and communities, though inequitably. The challenge of climate change is that it has created a shortfall in social work knowledge of and for practice that needs to be rectified. Through transformative learning, championed by educators in schools of social work as well as those in the field such new practice/s can be enacted. Such new practice/s ought to start with schools of social work creating the ideal conditions through information/ knowledge about the particular practice issues/demands which can help induce how practitioners' think about practice, attitudes toward such practice, and adaptiveness to such practice through recognizing old practice in new settings/contexts. Such was some of the thinking behind our decision to introduce the new learning area of environment and sustainability in our particular setting and to commit to formatively evaluating the initial outcomes to inform further development of the initiative. Mezirow (1996, 1997, 2000) and O'Sullivan et al. (2002) talk about transformative learning as relating to changes in attitudes, dispositions, and actions. Thus for students and practitioners to successfully enact practice using classroom knowledge, there is need to transform thinking and attitudes in ways that allow them to transition from mere attitude change to engagement among a group of learners with diverse perspectives (Mezirow, 1996).

The study findings demonstrate these transformations in various degrees for different participants. For some, it was *integrative* in enabling connections to be made between previously unrelated ideas and facts. Recommending 'thinking outside the square,' one student found it linked to 'all learning areas, not as a separate issue.' At the same time, many students and supervisors reported that this did not come easily to them and overtly sought support in moving into this new territory of understanding. The social work educational team also struggled as to how best to map this new territory, given the dangers of being too prescriptive in an area requiring open conversation and collaboration between all stakeholders. Through conducting the survey we sought to share facts, ideas, feelings, and opinions candidly while maintaining our focus on the educational objective of preparing students to practice eco-social justice and still maintain an effective dialogue, especially when emotions could run high.

Morgan's (2012) threshold concept characteristic of *boundedness* was represented in a variety of ways. Students stated their confusion in relating environmental sustainability to practice but were generally open to reconfiguring their learning. In contrast, four supervisors outrightly rejected the learning area for violating their sense of the boundedness of social work. In this, they were more forthright than students in articulating the *troublesome* nature of this eco-social work knowledge. The supervisor stating that no other university has such a learning area reflected a common cultural pattern of openly taking 'their cues from wider social and cultural depictions and scripts for understanding' (Morgan, 2012 p. 220). Students, given their relative powerlessness, were likely to navigate experiences of the troublesome nature of the topic by asking for specific practice instructions to be given.

Students having the agency to 'test the waters' in a supported environment and draw collaboratively on the knowledge of peers, supervisors, and most importantly clients are suggestions for the future curriculum development around Learning Area 8. It is expected that change to address this topic will occur incrementally over time. Reflection by the course team is clear that improvements could have been made on the first implementation of the initiative.

Many of the suggestions made through this survey have already been acted upon. At the same time in the wider world an increasing focus on responding to climate change also permeates the interactions and discussions between university academics, field supervisors, and students. The politics of climate change continue to add to the complexity and uncertainty of practice with the current Australian government implementing a significant change in direction from the previous government (Martin, 2015).

Through all this the course team considered it worthwhile to document their unilateral action in adding to the Practice Standards endorsed by the AASW and the results of an initial survey of students' and field supervisors' responses to this. This construction of change started small. As these efforts hone and amplify, they will join and resonate or differ from other initiatives addressing climate change in social work education. The final outcome in addressing this threshold topic is not known.

Conclusion

This paper has reported on a formative evaluation survey of student and field supervisor's responses to the inclusion of environmental sustainability as a learning area in field education. Responses have been connected to concepts guiding and scaffolding student learning drawn from educational research. In this, environmental sustainability is cast as a threshold concept where students and their supervisors explore liminal spaces of learning from a diverse range of positioning. With growing international support for eco-social work, such grounded local initiatives allow for formative evaluations of the transformative learning required for eco-social work practice. Clearly congruent are concerns to prepare social work graduates to address environmental sustainability and the experiential, critically reflective, and dialogical focus of transformative learning models. Challenges lie in teaching environment and sustainability content, including consciousness-raising and advocacy, while negotiating the perceptions of students and field educators as to the troublesome nature of this topic. The results of this survey of action taken in one rural Australian university suggest that it is possible to tackle emerging issues in social work in ways respectful of the complexities in crossing the threshold between conceptual thought and contextualized action.

Disclosure statement

No potential conflict of interest was reported by the authors.

Note

1. Savanna Sabine Agustine is now affiliated with Charles Sturt University, School of Humanities and Social Sciences.

References

Australian Association of Social Workers. (2013a). *Australian Social Work, 66*, 153–310.

Australian Association of Social Workers. (2013b). Practice Standards, retrieved from Australian Association of Social Workers. website: http://www.aasw.asn.au/document/item/4551

Bowers, C. A. (2009). Rethinking Social Justice Issues Within an Eco-Justice Conceptual and Moral Framework. *Journal of Educational Controversy, 4*. Retrieved February 20, from: www.wce.wwu.edu/resources/cep/ejournal/v004n001/a006.shtml .

Coates, J., & Gray, M. (2012). The environment and social work: An overview and introduction. *International Journal of Social Welfare, 21*, 230–238. doi:10.1111/j.1468-2397.2011.00851.x

Cousin, G. (2006). An introduction to threshold concepts. *Planet, 17*, 4–5. Retrieved from http://www.et.kent.edu/fpdc-db/files/DD%2002-threshold.pdf

Dominelli, L. (2011). Climate change: social worker's roles and contributions to policy debates and interventions. *International Journal of Social Welfare, 20*, 430–438. doi: 10.1111/j.1468-2397.2011.00795.x

Eco-Social Work: Transforming the national social work curriculum. (2013). (conference), Brisbane, Australia.

Gibson, C., Waitt, G., Head, L., & Gill, N. (2011). Is it easy being green? On the dilemmas of material cultures of household sustainability. In R. Lane & A. W. Gordon-Murray (Eds.), *Material Geographies of Household Sustainability* (pp. 19–33). Surrey, England: Ashgate Publishing Limited.

Hannam, P. (2014 November 26). "Climate variability" will hit NSW's water security, government says. *The Sydney Morning Herald, 5*. Retrieved from http://www.smh.com.au

Jones, P. (2011). Transformative Learning, Eco Pedagogy and Learning Ecology: Pathways for Social Work Education. In P. Jones, D. Miles, A. Francis, & S. P Rajeev (Eds.), *Eco-social justice: Issues, challenges and ways forward*. Karnataka India: Books for Change Publishers.

Martin, P. (2015 January 11). Retrieved from http://www.smh.com.au Climate change: Why some of us won't believe it's getting hotter. *The Sydney Morning Herald.*

Maple, M., Jarrott, H., & Kuyini, A. B. (2013). Blended learning in rural social work education: Reflections from a new Australian bachelor of social work course. *Social Work Education, 32*, 349–364. doi:10.1080/02615479.2012.668181

McKinnon, J. (2013). The Environment: A private concern or a Professional practice issue for Australian Social workers. *Australian Social Work, 66*, 156–170. doi:10.1080/0312407X.2013.782558

Meyer, J., & Land, R. (2003). Threshold Concepts and Troublesome Knowledge: Linkages to Ways of Thinking and Practising within the Disciplines, Enhancing Teaching – Learning Environments in Undergraduate Courses. *(Project Occasional report 4)* (pp. 1–14). Retrieved from: http://www.etl.tla.ed.ac.uk//docs/ETLreport4.pdf

Mezirow, J. (2000). Learning to think like an adult: Core concepts of transformation theory. In J. Mezirow & Associates (Ed.), *Learning as transformation: Critical perspectives on a theory in progress* (pp. 3–34). San Francisco, CA: Jossey-Bass.

Mezirow, J. (1996). Contemporary Paradigms of Learning. *Adult Education Quarterly, 46*, 158–173.

Mezirow, J. (1997). Transformative Learning: Theory to Practice. *New Directions for Adult and Continuing Education, 74*, 5–12.

Molyneux, R. (2010). The Practical Realities of Ecosocial Work: A Review of the Literature. *Critical Social Work, 11*, 61–69. Retrieved from http://www.uwindsor.ca/criticalsocialwork/the-practical-realities-of-ecosocial-work-a-review-of-the-literature

Morgan, H. (2012). The social model of disability as a threshold concept: Troublesome knowledge and liminal spaces in social work education. *Social Work Education, 31*, 215–226. doi:10.1080/02615479.2012.644964

Norton, C. (2012). Social work and the environment: An ecosocial approach. *International Journal of Social Welfare, 21*, 299–308. doi:10.1111/j.1468-2397.2011.00853.x

O'Sullivan, E. V., Morrell, A., & O'Connor, M. A. (2002). *Expanding the boundaries of transformative learning.* New York: Palgrave.

Patton, M. (2008). *Utilization-focused evaluation* (4th ed.). Thousand Oaks, CA: Sage.

Schwandt, T. (2014). On the mutually informing relationship between practice and theory in evaluation. *American Journal of Evaluation, 35*, 231–236. doi:10.1177/1098214013503703

Thompson, N. (2002). Social movements, social justice and social work. *British journal of social work, 32*, 711–722. doi:10.1111/j.1468-2397.2011.00853.x

Zapf, M. (2010). Social Work and the Environment: Understanding people and place. *Critical Social Work, 11*, 30.

Reflexive Development: A Model for Helping Social Workers Contribute to a Sustainable Global Future

Eric A. Des Marais, Sarah M. Bexell & Subhasis Bhadra

It is difficult for many social work students to grasp the importance they and their traditional client base play in global environmental issues. International development generally focuses on supporting human development through the development of natural resources. However, increased human consumption everywhere in the world is disrupting the global system upon which all life depends. Continued Western consumption habits are simply unsustainable. It is therefore incumbent that any increase in consumption, even in the 'developing' world, be considered in terms of global environmental sustainability; change must also occur within the developed world. To address this problem, reflexive processes that support transnational analysis and action must be developed. This paper suggests classroom activities that help students analyze and problem solve around this process of Reflexive Development.

Introduction

The conceptual approach of person-in-environment has long been a mainstay in the field of social work in the United States (Germain, 1973, 1978). While it is an excellent conceptualization for social workers focused on the individual level, social work students can have difficulty using it to account for social and environmental properties that only emerge at the community, national, or global level (McKinnon, 2008). For social work students, it can be especially difficult to grasp those complex, transnational problems such as climate change and the resulting effects on local systems such as increased natural disasters are more than the mere sum of individual actions.

Similarly, students have difficulty grasping how consumption in the developed world bears direct impact upon the developing world. The field of social work has an important role to play in the fields of sustainable development and humane education, and social work students need a conceptual framework for learning about it (Mary, 2008). Traditionally, sustainable development refers to the utilization of natural resources to support human development while maintaining the local environment for future generations (Brundtland Commission, 1987). Building on the methodological approach of humane education and the concept of reflexivity, a core skill of social work practice, we suggest that students can learn to grasp these transnational processes and learn how to affect change on them through Reflexive Development.

While Reflexive Development has in the past been defined as incorporating reflexivity into the social processes of international development (Pieterse, 1998), we suggest that Reflexive Development is the process of nurturing or growing transnational reflexivity in which the process of social and ecological change in which human life choices in developed countries are responsive to the needs of developing areas of the world, but also maintain sustainability. In international development, development usually refers to the process of helping people in poor countries attain better lives. As increased consumption in rich countries puts pressure on developing countries, change must also occur within developed countries—part of the development process in rich countries must be to reduce waste and overconsumption, because it is having negative consequences on the security of other people, global climate, and all Earth's systems. The concept of Reflexive Development helps to not only critique current approaches to development, but also to point the way to novel solutions.

Ecological Boundaries

For survival, humans depend completely on Earth's ecosystems and the services they provide, such as clean air and water, food, medicines, disease management, building materials, fuel, climate regulation, buffering of storms, spiritual fulfillment, and aesthetic enjoyment. Over the past 50 years, humans have negatively altered global ecosystems more rapidly and extensively than in any comparable period of time in human history (WWF, 2014). The vast and detrimental changes to Earth's ecosystems are largely to meet rapidly growing, unsustainable and irresponsible human demands for food, fresh water, timber, fiber, minerals, and fuel (Global Humanitarian Forum, 2009; Millennium Ecosystem Assessment, 2005b). This transformation of our planet has contributed to substantial gains in human well-being and economic development for a few, but at the cost of most human populations as well as almost all other species we share the Earth with. The global ecological costs associated with these gains of a few people have only recently become apparent. We now know that we as a species have so degraded and destroyed parts of our life-support system that our health, well-being, and survival are threatened (Raworth, 2012; Rockström et al., 2009; Zapf, 2010). It is increasingly clear that environmental problems cannot be resolved through technical solutions alone and that these ecological issues are forcing us to rethink what development and progress means. These threats are not as new as some might think, with one of the most prominent calls to action coming from the *Limits to Growth* studies, first published in 1972 (Meadows, Randers, & Meadows, 2004).

It would be prudent to consider our conscious and unconscious neglect and abuse of the environment as a widespread social disorder and explore the consequences for failure to understand that human behavior has created and continues to accelerate the decline of global biodiversity that imperils the existence of most species, including humans (Rockström et al., 2009; Tedeschi, Bexell, & Nesmith, 2013; Zapf, 2010). While several authors attempted to take note of the environment as essential to the consideration of human health (e.g., Lehmann & Coady, 2001; Miley, O'Melia, & DuBois, 2004; Sheafor & Horejsi, 2006; Yelaja, 1985), and several devoting their entire career to enlightening the profession (e.g., Fred Besthorn, John Coates, Mel Gray), their suggestions have gone unheeded and as the world watches daily news of environmental deterioration, we remain ineffective in helping the very humans we signed an oath to protect. What are the manifestations of our human-centric view of the world and avoidant practices (i.e. denial) on human physical and psychological health? To say the field of social work is aware but then not take action is problematic.

Linked Social and Ecological Systems

Threats of climate change, mass extinction, water shortages, limits of non-renewable resources, and more prompt us to think holistically and develop the understanding of linked social and ecological systems. Traditionally, social and ecological systems were considered separate and even analyzed by different types of scientists and professionals. This reductionist approach was also prevalent in the early roots of social work (Zapf, 2010). Today we recognize the importance of scientists from a broad array of disciplines working to counteract past shortsightedness (Manfredo, Vaske, Rechkemmer, & Duke, 2014). One current perspective is that Social Ecological System (SES) include ecological systems that are intricately linked with and affected by one or more social systems (Anderies, Janssen, & Ostrom, 2004). Ecological systems are loosely defined as interdependent systems of biological units and geophysical elements, while social systems as interdependent systems of social organisms. Social and ecological systems possess units that interact interdependently. Because social and ecological systems are interlinked, an SES is a complex, adaptive system involving multiple subsystems, as well as being embedded in multiple larger systems (Anderies et al., 2004). To disregard one unit is to threaten others.

Earth's continued sustainability for humans can no longer be taken for granted (Millennium Ecosystem Assessment, 2005a; Rockström et al., 2009; Steffen et al., 2015). We must first acknowledge our dismissal of ecological parameters as important to the field of social work. Environmental change due to unchecked human population growth and overconsumption now threaten our survival, with the poorest and most vulnerable on Earth impacted most significantly (Raworth, 2012), making this a social justice issue facing social work practitioners.

Humans living in developed countries have been initially buffered against environmental changes by technology, the very foundation of global climate disruptions that are especially affecting millions of people living in developing countries (Global Humanitarian Forum, 2009). However, even those living in developed countries are now experiencing the effects of global decline (e.g., increased severity and frequency of natural disasters, mass environmental immigration) and our fundamental and inextricable dependence on nature's services.

Anthropogenic-induced changes to Earth's systems have caught up to humanity, forcing us to reassess our activities in order to insure that our planet remains habitable for us (Steffen et al., 2015). In the context of global human 'development' this has become a paramount consideration as 'developed' nations must drastically decrease their ecological footprint in short-order to afford sustainable use of natural resources and begin to reverse the negative consequences of global climate change, lift nearly two billion people out of extreme poverty, and create a social paradigm that respects the importance of other living beings and protection of biodiversity, of which we have lost 52 percent (not in total number of species but in total volume) in just the past 40 years (WWF, 2014).

Myth of Infinite Economic Growth

Misguided perceptions of what constitutes human 'progress' has resulted in humans living for material gain versus proper respect, appreciation, and compassion for each other as well as all other beings. Many humans disregard or even have an intentionally destructive attitude toward other life forms, those that provide the environments allowing for life to exist on our planet at all. Unfounded beliefs of superiority have fueled our disregard for the significance of our own population size that, coupled with consumption, impose limits on human well-being. Because material gain has nearly unanimously been used as our measure of progress and even happiness, the myth of infinite economic growth has fueled economies that are unsustainable and on the verge of collapse. We live on a finite planet with finite resources, therefore when every nation on Earth measures its success on an increasing GDP we are behaving illogically and are attempting to defy the very laws of physics by which all life must conform. As the ecological economist Daly (1988) shared, 'it would be wise to think of the economy as a subsystem of an ecosystem that is energetically finite and in accord with physical laws must stop growing' (p. 13). However, way before this point was reached, human overconsumption had already caused rampant destruction of global ecosystems and some have argued, so far breached carrying capacity (the point at which populations exceed the environments' capacity to regenerate itself) that some ecosystems may never return to their pre-destruction state. This misunderstanding of ecological limits and incomplete economic models has created the paradox of economic development causing ecological disaster that will eventually harm everyone and possibly lead to human extinction (Yule, Fournier, & Hindmarsh, 2013) or at the very least, drive the Earth system into a much less hospitable state for humanity (Steffen et al., 2015).

Social Work and Its Struggle with Environmental Justice

Fundamental to equipping social workers with the knowledge and skills to be effective in all areas of the human condition is the inclusion of a comprehensive understanding of the critical importance of healthy environments, ecological boundaries, and the environmental forces that create, contribute to, and address problems in the everyday lives of people. Social work as a discipline has useful models and language referencing the concept of *person-in-environment*, but has avoided integrating these concepts to include more than human-centric, social consideration (Mary, 2008; Tedeschi et al., 2013; Zapf, 2010). Kemp (2011) identifies the person-environment approach as one of three conceptual frameworks to incorporate environmental concerns into social work.

The person-in-environment approach is considered at the core of social work education, and it 'highlights the importance of understanding an individual and individual behavior in light of the environmental contexts in which that person lives and acts' (Kondrat, 2013). Students often understand it or visualize it in terms of the Human Ecological Theory of Bronfenbrenner (1979), a model that accounts for the multi-level systemic factors that affect child development, or as layers of an onion, much like the box in boxes metaphor of Germain (1978). It is well suited for helping clinical social workers and case managers conceptualize the impact of social structures such as families, communities, and social policies on their clients, and for how their clients can in turn impact those social systems. However, it is strictly a psychological/social theory, and as such, it makes a clear (and we would argue, false) delineation between social systems and ecological systems (Zapf, 2010). What is more, it also fails to account for the emergent properties of complex interactions that occur within SESs—*not* merely the ecology of social systems social workers are already familiar with, but with the feedback loops between social systems and the systems of the physical environment upon which life depends (Walker & Salt, 2006).

Despite its usefulness in helping social work students appreciate the affects of larger systems on individual functioning, when misapplied, the person-in-environment approach makes it particularly difficult for social work students to grasp the importance of environmental issues. Many social work students only see themselves as case managers or clinicians working one-on-one with a client, and it can be particularly hard to help them understand why environmental issues, especially issues such as climate change, are a necessary and perhaps even fundamental concern for social workers.

It has been argued that the practice of social work does not extend beyond the bounds of the social. For example, in Kondrat's (1999, 2003) notable constructivist conceptualizations of the social and the personal that incorporate macro processes more fully, the influence of and dependence upon the natural environment are not accounted for. However, as Gray, Coates, and Hetherington (2012) illustrate, social workers have played and continue to play an important role in the environmental movement. It therefore is incumbent upon social work educators to help students grasp the necessity of social workers in sustainability. Kemp (2011) outlines a range of issues that social work struggles with in terms of the environment.

Social work students are often focused on learning to work with individuals in their own communities, and can feel that whatever positive role they play in an individual's life adds to the social justice in the world. However, when they are not taught to look beyond the bounds of their local social systems, they cannot recognize important transnational factors that do directly affect their local client systems. For example, in developed countries economic materialism, a value system in which status and happiness is measured by the accumulation of material wealth, is the de facto measure of worth (Richins, 1994). It is this very materialism, though, that strains ecological systems to the breaking point, and creates systems of injustice such as the rise in modern forced labor. The quest for obtaining equality for individual clients in one locale (increasing consumption in the United States) can actually reinforce emergent feedback loops that propagate environmental injustice.

This issue is not just relevant to social workers in the developed world, but also to those in developing and underdeveloped countries. For example, India is on the path to westernization and modernity, including better education, health services, and livelihood development. The influence of globalization and technology cannot be denied, yet it has not filtered down to all communities. Marginalized geographically (living in remote villages), educationally (illiterate), or through lack of accessibility to basic services (health care facilities, communications), these communities are exploited by the same development processes that others in the country are benefiting from. The very natural resources indigenous people globally have relied on for generations are exploited for the consumerism of others. Industrialization and urbanization within India creates the same exploitation that industrialization and urbanization in Western countries have caused. For example, the Himalayan Tsunami of 2013 created severe landslides due to poorly planned road construction and housing development and these were the very development processes that magnified the disaster leading to disrupted access to natural resources for local people. Mitigation of disaster risk was weakened by the construction of multistoried housing on hills, poor drainage along new roads, and deforestation. The promotion of tourism and other economic interests took precedence over sustainability of resource ultilization. After the disaster, local people suffered the worst. Except for a few local NGOs, the issue of environmental justice was never addressed, neither by the state nor national governments.

The above problems are intertwined, and the answer for managing all of them together is not yet clear; this can be challenging for students. One key aspect to adding environmental issues into the psyche of social workers is give them a model to work from that helps them to not just conceptualize and critique these issues, but help them identify potential solutions as well, especially those in their own locales. By providing this model, social workers can then work towards building a more sustainable world, both locally and internationally. Our approach is closely aligned with the globally focused category in Kemp (2011) taxonomy of conceptual approaches to environmental issues. In this framework, the focus is on the unjust impact of global environmental degradation on at risk populations.

Reflexive Development

As a theoretical framework, we begin from Luhmann's conceptualization of social systems. While systems theories are often criticized for being overly focused on homeostasis, that is, on how systems don't change, his conceptions of reflexivity and reflection are particularly useful for our endeavor. First, for Luhmann (1993), social systems refer to systems of, and only of, communication. While this boundary could appear problematic in terms of our goal of linking social and ecological systems, it is important to note that he doesn't argue that other systems don't exist, but that social systems are systems of communication—the exchange of information. A social system is different from its environment in that it is an area of less complexity than its environment (Luhmann, 1995). A system is also different from other systems because it has its own communicative processes (Luhmann, 1995). His theory is especially problematic for the person-in-environment approach because it examines communication as existing apart from the individual, thus seemingly removing the social from the individual. While this is a drawback of the theory in terms of working directly with clients (Gershon, 2005), it helps to refocus on social processes in and of themselves.

Another facet of Luhmann's perspective that is important for helping students is that of observation. For Luhmann, a system can both engage in a "first order" observation of itself, which he calls a *reflexive* process, and it can also observe these first order processes, which he calls *reflective* process (Luhmann, 1993, 1995). As an example, when social workers are engaging in their daily work processes such as helping clients navigate their communities, they are engaging in a first order communicative or reflexive process, but when they examine how this process interacts with its environment or other systems, then they are engaging in a reflective process.

Coming from this standpoint, we believe that it is necessary to develop transnational reflexive processes and that the first step in doing so is engaging in reflection about what is happening globally. Thus, students need to first observe how their immediate lives and the lives of their clients/client systems interact with other systems. Once the process of reflection occurs, it is then possible to help new systems couple together more effectively (for example, environmental professionals in over-developed countries work with indigenous populations in under-developed countries, or more importantly within their own) or to support the development of new transnational systems (such as international systems of sustainable development and environmental governance). The weakness of our approach, however, is that Luhmann (1993, 1995) only addresses social processes and not natural environment systems, and this needs to be reincorporated.

So, within this approach, Reflexive Development is a process of transnational social and ecological change in which human life choices in developed countries become responsive to the needs of developing areas and to the planet as an integrated social ecological system. It suggests that we need new ways to do the following: (1) transmit knowledge of how development in one area of the world affects development in another area of the world; (2) increase sustainable use where human flourishing does not exist; and (3) limit consumption that harms other people and other species; (4) limit consumption that harms ecosystems; and (5) transition human thinking about what constitutes happiness and well-being. The use of the term 'reflexive' is especially important, as it is a concept that social work education already utilizes and is thus comfortable for both students and faculty.

Our use of the term *Reflexive Development* is related to Pieterse (1998) usage of the term, who argues, 'development has been reflexive all along in that development ideas and policies have been reactions to and reflections on the failures and limitations of previous development practices' (p. 368). Pieterse (1998) also argues that the process of international development involved multiple layers of reflexivity, including local, community, and national levels. What we suggest is that social workers need to be part of a conscious process of developing transnational reflexivity. So while, Pieterse sees international development as a process that grows through reflexive recursion, our perspective is that we must develop reflexive processes on a transnational level.

Reflexivity in Social Work Education

Reflexivity has been used in various social perspectives to explain the cyclic impacts of events in a 'cause and effect' relationship in different contexts. Social perspective includes the wider socio, political, economic, cultural perspective that determines the well-being and existence of an individual, family and community. Human action is having a purpose in relation to the possession of the action and rationalization of the action. This process produces self-reflection. Thus, individual reflexivity is being understood not merely as self-consciousness but as the mental monitoring of the flow of the social life. Human action (i.e. response) is not an automated one, but rather internally characterized acts that agents believed would have a particular outcome (Peet & Hartwick, 2005). Reflexivity has an important role in understanding the development paradigm and in analyzing the sequence of development in a context of environmental sustainability and its relevance for social work practice.

Social work as a profession is based on the development of professional-self and professional ability that inculcates values, knowledge and skills. Thus, for a social worker, use of her/his professional self becomes pivotal for practice. In the development of this professional-self the methods of social work teaching play the key determinant role. Social work theory, education and practice are interconnected with learning and self- reflection. Thus the term reflexivity for a social worker is a combination of sensitivity, identification or self-observation and self-presentation (Holmström & Kjaerbeck, 2007). Reflexivity in social work literature is also being used as 'critical reflection'. D'Cruz, Gillingham, and Melendez (2007) explained reflexivity for social work practice in three dimensions. The first is reflexivity as an individual's considered response to an immediate context and making choices for further direction. The second dimension of reflexivity is an individual's self-critical approach. The final dimension is about role of emotion in social work practice.

We suggest that students need to appreciate not only individual level reflexivity, but also systems-level reflexivity, in which reflexivity is about one sub-system causing and/or responding to change in another part of the system. Through a process of gaining new knowledge about disparate parts of the global Social Ecological System, social workers are then able to participate in the creation of just and ecological sustainable transnational systems of development. Thus, Reflexive Development is the nurturing of transnational systems that not only help push forward the advancement of areas of the world struck with poverty, disease, and oppression, but that also work within developed countries to reduce the overconsumption that induces and propagates ecological destruction and social injustice.

We have designed activities that actively engage our own students in reflecting upon their own place in the global ecosystem, as well as the place of their clients in relation to other client systems. This helps students to better grasp how trying to manage social change in their own country can have unintended consequences, both social and ecologically, in other parts of the world. However, we also try to move beyond the critical approach, and this is where our use of Reflexive Development is particularly useful—it gives students a conceptual framework for moving beyond critical social work and into creative social work by helping to examine current attempts to build transnational reflexive processes. These activities help students to examine what tools and social systems are currently being used to engage transnationally, and what work is still to be done.

The following activities have all been developed and used in our own social work classes. The classes are within a sustainable development concentration that utilizes a perspective of coupled social and ecological systems. In the classes of the concentration, students develop an understanding of Social Ecological Systems, the human dimensions of climate change, and the role of social workers in the field of international development.

The Three Dimensions in the Classroom

The first dimension of reflexivity refers to a social workers' ability to gain and create knowledge that guides life choices (D'Cruz et al., 2007). Reflexivity becomes a competency or ability to act in the world consciously, and thus to become an active citizen, and in our sense, a *globally* active citizen (Ferguson, 2003). In terms of Reflexive Development, this means social work students need to first begin to understand what is happening in other parts of the world and how these events are linked to their own lives, the lives of their clients, and their communities.

Our students have a varying degree of understanding of global environmental issues, its causes, and how it affects not just their own lives, but also those of their clients, and people in other parts of the world. In order to help them grasp the gravity of climate change, we begin by giving students appropriate reading such as Raworth (2012), which utilizes the image of a doughnut as a metaphor for safe and just use of resources on a global level. It explains both global inequality and sustainability in way that is engaging and critical while also leaving the door open for what solutions might have to look like.

Following the presentation, small group work with students generally seems to have them grasp the importance of sustainability, but to not yet recognize how they fit into it. This leads to the next step, focusing on self-criticality.

Self-criticality is a key step in helping students grasp that development in their own country, and their own consumption and child bearing, has direct implications for individuals in other countries, as well as their own. There are a variety of online 'footprint' calculators that help students understand the effects their own consumption has globally.

One of the first we used, although not focused on sustainability, helped students tally their possessions and determine how many slaves are in the supply chain of their possessions (www.slaveryfootprint.org). Another one, The Global Footprint Calculator (http://footprint.wwf.org.uk), helps students calculate their carbon footprint and then tells them how many Earths would be required if everyone consumed as much as they did A student picks their location and answers questions about their consumption of food, goods, shelter, and mobility. Then a score is calculated showing how many planets would be required to provide that level of usage if everyone on the planet consumed the same way. Students can also calculate a hypothetical client's consumption and then recalculate for the same client after they receive services to see how that consumptive change would affect sustainability. This helps students understand that a narrow approach to helping individuals can actually feed larger feedback loops of injustice. These two websites also provide methodological notes, so there can be a discussion of research methodologies as well. After using the calculators, we have students engage in class discussions about what this means in terms of social justice. One discussion point that stirs discussion greatly is bringing forward the question of whether helping one client system by shifting the burden of injustice to someone else truly fulfilled our ethical responsibility of creating social justice.

Moving on from the global ecosystem and also helping to bridge to the final dimension of reflexivity, it is important to help students examine their own beliefs and emotions around non-human forms of life. Especially if the student comes from a strong Judeo-Christian background, it can be difficult to get past the assumption of human primacy. Exercises here involve readings, videos, and discussions on parental and other social and intellectual activities in other animal species and then comparison to human activities. Useful and engaging questions that have stirred our students involve reframing discussions such as, 'Do chimpanzee mothers feel different than human mothers when their children are hurt? What about dog mothers?' This is then followed with discussions of whether humans are that different from other animals, and are human and other forms of life in competition or in cooperation.

To get students to engage emotionally with non-human life, creative writing exercises either in class or at home from the perspective of nonhuman actors such as other animals, trees, amoebas, or even bacteria is a fun way to explore life relationships. Having the students share the stories with each other and reflect on what it was like to take the perspective of a nonhuman can be helpful to cement the connection.

Another exercise is to have students think about the 'footprint' of any other individual of a species on Earth and guide them to realize that other species do not have a negative footprint (unless they are an invasive, which is due to human error, or all their predators have been killed off by humans) and in fact their activities provide positive feedback loops for the health of the environment. Then ask students to think about what our modern human activities give back to the environment (pollution, waste, depletion). This activity is designed to inspire respect for other life forms and a greater sense of urgency to live more in harmony with nature.

While helping students conceptualize and internalize models of translocal effects (activities in one locality influencing the functioning of another locality) is important, it is just as important to move to the next step as well—action. Leaving students with a critique is only half of the battle. Helping them creatively engage with the issues that are presented through a Reflexive Development analysis to consider solutions is also important, and will steer them away from apathy or hopelessness.

Pointing the Way Forward

Having students investigate people, organizations, and activities that support Reflexive Development helps them to engage in moving beyond critique into solution-making. Identifying both local and international actors is an important step to help social workers find their place in the environmental justice movement.

Bringing in presenters who focus on urban or organic farming, sustainable design, or city planning can help students understand the local actors involved in sustainable initiatives. A relatively famous author, presenter, and organizer that resonates particularly well with students is Vandana Shiva. Shiva exemplifies many social work values in an environmental framework. Her books (e.g., Shiva, 2006) are appealing and her presentations (many of which are available on YouTube) are thoughtful and engaging.

Service learning activities also help students become involved. Local river clean ups, bird counts, urban gardens, and even volunteering with larger groups such as World Wildlife Fund, Audubon, or Greenpeace, can be great opportunities to help students get direct experience with supporting environmental justice. Also, these organizations rarely get to interact with people with the community and systems skills that social workers possess, therefore the benefits will surely be mutual.

Instilling a stronger understanding of what sustainability is and looks like is also constructive. Sustainability doesn't mean zero growth, but rather new ways of thinking about growth, about qualitative growth, not physical expansion of the human machine (Meadows et al., 2004). A sustainable society would consider material goods as a tool, but not a perpetual mandate and begin to discriminate among kinds of growth and purposes for growth. It would also consider the true costs of any kind of growth, who would actually benefit how long it would last and whether that growth could be accommodated by the sources and pollution sinks of Earth. Lastly, it would consider types of positive social growth that would include deeper empathy with each other, Earth and other beings, quality time spent with family and friends in meaningful activities, and access to meaningful work and play that supports society and is not just a drain such as the economic machine and policies now dominant worldwide. When teaching our students about these issues, they often go through a natural cycle of denial, then guilt, then thoughts about how to make personal and professional changes, through to hope and determination to be a part of the challenge and adventure being presented to their generation. Our guidance, love and support is paramount.

Finally, to help grasp the transnational approaches to environmental justice, we help students research a variety of topics. Students look for smartphone apps such as the *Environmental Justice Participatory* app or the *How is the Water?* that support environmental justice. Some have also joined the *Slow Food* and *Slow Fashion* movements. They can also examine online attempts to build transnational civil society through websites such as Avaaz and CIVICUS. New payment methods such, as Bitcoin (https://bitcoin.org/en/) that bypass traditional banking are another new area of transnational connection. Students can also examine organizations of international governance such as the United Nations Development Program, the Sustainable Development Solutions Network, and the United Nations Office of Danger Risk Reduction. Examining their job boards can help students understand how their own skill sets can fit into the field.

Conclusion

Models are simplifications of the real world that help us to better analyze and act upon that world (Meadows, 2008). Reflexive Development is a model that helps social workers to conceptualize, analyze, and act upon global social ecological systems in a way that supports environmental justice as well as more traditional sustainable development. Since the consumption of resources in one locality is tied to issues of justice in other locales (translocality), these translocal systems must be taken into account when social workers are acting to support social and environmental justice. Far from being a simple matter of helping developing areas of the world reach the economic level of the developed nations, it is necessary to change patterns of overdevelopment in the developed countries to also support sustainability. Thus, Reflexive Development recognizes that there is a dual process of supporting development and mitigating overdevelopment. By helping social work students better understand how their own participation and modeling is a necessary component of sustainability social work can play its role in supporting environmental justice.

While the purpose of this paper is to present a model for helping social work students grasp the importance of transnational networks that link localities and then act on them, local issues themselves should not be forgotten. In fact, getting students involved in local environmental issues will build the bridge to understanding larger global processes.

For example, asking students to think about the full impact of the development of indoor plumbing that then depletes a local aquifer shows students that things we take for granted and believe are necessary for well-being can have dangerous impacts and need to be considered wisely. Then linking this localized water issue to transnational issues, such as water wars in Africa, helps students appreciate that their local struggles also have wider implications in the world. Another example is the building of an oil pipeline in the name of national security, but that the pipeline polluting the water and soil impacts the people and other animals that live along the route. These issues are far from easy to resolve, but getting social workers at these tables, asking the hard questions, and above all advocating for the rights and well being of all impacted will surely bring about a more just and healthy human presence on Earth.

Disclosure statement

No potential conflict of interest was reported by the authors.

References

Anderies, J. M., Janssen, M. A., & Ostrom, E. (2004). A framework to analyze the robustness of social-ecological systems from an institutional perspective. *Ecology and Society, 9*, 18–34.

Bronfenbrenner, U. (1979). Contexts of child rearing: Problems and prospects. *American Psychologist, 34*, 844–850. doi:10.1037/0003-066X.34.10.844

Brundtland Commission. (1987). *Report of the world commission on environment and development.* Oxford, UK: Oxford University Press.

D'Cruz, H., Gillingham, P., & Melendez, S. (2007). Reflexivity, its meanings and relevance for social work: A critical review of the literature. *British Journal of Social Work, 37*, 73–90. doi:10.1093/bjsw/bcl001

Daly, H. E. (1988). *Beyond growth.* Boston, MA: Beacon Press.

Ferguson, H. (2003). Welfare, social exclusion and reflexivity: The case of child and woman protection. *Journal of Social Policy, 32*, 199–216. doi:10.1017/S0047279402006967

Germain, C. B. (1973). An ecological perspective in casework practice. *Social Casework, 54*, 323–330.

Germain, C. B. (1978). General-systems theory and ego psychology: An ecological perspective. *Social Service Review, 52*, 535–550. doi:10.1086/643676

Gershon, I. (2005). Seeing like a system: Luhmann for anthropologists. *Anthropological Theory, 5*, 99–116. doi:10.1177/1463499605053993

Global Humanitarian Forum. (2009). *Human impact report: Climate change—The anatomy of a silent crisis.* Geneva: Author.

Gray, M., Coates, J., & Hetherington, T. (Eds.). (2012). *Environmental social work.* New York, NY: Routledge.

Holmström, S., & Kjaerbeck, S. (2007). The reflective paradigm, in theory and practice- an inter disciplinary practice. In *Conference legitimacy in a changing world.* Roskilde: Roskilde University. Retrieved from udar.ruc.dk/bitstream/../The_reflective_paradigm__konferenceartikel.d.

Kemp, S. P. (2011). Recentring environment in social work practice: Necessity, opportunity, challenge. *British Journal of Social Work, 41*, 1198–1210. doi:10.1093/bjsw/bcr119

Kondrat, M. E. (1999). Who is the "self" in self-aware: Professional self-awareness from a critical theory perspective. *Social Service Review, 73*, 451–477.

Kondrat, M. E. (2003). Actor-centered social work: Re-visioning "person-in-environment" through a critical theory lens. *Social Work, 47*, 435–448.

Kondrat, M. (2013, June 11). Person-in-environment. *Encyclopedia of Social Work.* Retrieved from http://socialwork.oxfordre.com/view/10.1093/acrefore/9780199975839.001.0001/acrefore-9780199975839-e-285

Lehmann, P., & Coady, N. (2001). *Theoretical perspectives for direct social work practice: A generalist-eclectic approach.* New York, NY: Springer.

Luhmann, N. (1993). *Risk: A sociological theory.* New York, NY: Walter de Gruyter.

Luhmann, N. (1995). *Social systems.* Stanford, CA: Stanford University Press.

Manfredo, M. J., Vaske, J. J., Rechkemmer, A., & Duke, E. A. (2014). *Understanding society and natural resources: Forging new strands of integration across the social sciences*. New York, NY: Springer Open.

Mary, N. L. (2008). *Social work in a sustainable world*. Chicago, IL: Lyceum Books.

McKinnon, J. (2008). Exploring the nexus between social work and the environment. *Australian Social Work, 61*, 256–268. doi:10.1080/03124070802178275

Meadows, D. H. (2008). *Thinking in systems: A primer*. White River Junction, VT: Chelsea Green Publishing.

Meadows, D., Randers, J., & Meadows, D. (2004). *Limits to growth: The 30-year update*. White River Junction, VT: Chelsea Green Publishing.

Miley, K. K., O'Melia, M., & DuBois, B. (2004). *Generalist social work practice: An empowering approach* (4th ed.). Boston, MA: Pearson Education.

Millennium Ecosystem Assessment. (2005a). *Ecosystems and human well-being: Biodiversity synthesis*. Washington, DC: World Resources Institute.

Millennium Ecosystem Assessment. (2005b). *Ecosystems and human well-being: Synthesis*. Washington, DC: Island Press.

Peet, R., & Hartwick, E. (2005). *Theories of development*. New Delhi: Rawat Publication. (Indian Reprint).

Pieterse, J. N. (1998). My paradigm or yours? Alternative development, post-development, reflexive development. *Development and Change, 29*, 343–373.

Raworth, K. (2012). A safe and just space for humanity: Can we live within the doughnut? Oxfam Discussion Papers. Retrieved from http://www.oxfam.org/sites/www.oxfam.org/files/dp-a-safe-and-just-space-for-humanity-130212-en.pdf

Richins, M. L. (1994). Valuing things: The public and private meanings of possessions. *Journal of Consumer Research, 21*, 504–521. doi:10.1086/209414

Rockström, J., Steffen, W., Noone, K., Persson, Å., Chapin, F. S., Lambin, E. F., . . . Foley, J. A. (2009). A safe operating space for humanity. *Nature, 461*, 472–475. doi:10.1038/461472a

Sheafor, B. W., & Horejsi, C. R. (2006). *Techniques and guidelines for social work practice* (7th ed.). Boston, MA: Pearson Education.

Shiva, V. (2006). *Earth democracy: Justice, sustainability and peace*. Cambridge, MA: South End Press.

Steffen, W., Richardson, K., Rockström, J., Cornell, S. E., Fetzer, I., Bennett, E. M., . . . Sörlin, S. (2015). Planetary boundaries: Guiding human development on a changing planet. *Science, 34*. doi:10.1126/science.1259855.

Tedeschi, P, Bexell, S. M., & Nesmith, J. (2013). Conservation social work: The interconnectedness of biodiversity health and human resilience. In M. Bekoff (Ed.), *Ignoring nature no more: The case for compassionate conservation* (pp. 223–235). Chicago, IL: University of Chicago Press.

Walker, B., & Salt, D. (2006). *Resilience thinking: Sustaining ecosystems and people in a changing world*. Washington, DC: Island Press.

WWF. (2014). *Living planet report 2014: Species and spaces, people and places*. Gland: Author.

Yelaja, S. A. (1985). Concepts of social work practice. In S. A. Yelaja (Ed.), *An introduction to social work practice in Canada*. Scarborough, ON: Prentice Hall.

Yule, J. V., Fournier, R. J., & Hindmarsh, P. L. (2013). Biodiversity, extinction, and humanity's future: The ecological and evolutionary consequences of human population and resource use. *Humanities, 2*, 147–159.

Zapf, M. K. (2010). Social work and the environment: Understanding people and place. *Critical Social Work, 11*, 30–46.

Index

For Product Safety Concerns and Information please contact our EU
representative GPSR@taylorandfrancis.com
Taylor & Francis Verlag GmbH, Kaufingerstraße 24, 80331 München, Germany

www.ingramcontent.com/pod-product-compliance
Ingram Content Group UK Ltd.
Pitfield, Milton Keynes, MK11 3LW, UK
UKHW050753180926
475701UK00007B/1030
* 9 7 8 1 1 3 8 3 9 2 0 9 0 *